Patrick Moore's Practical Astronomy Series

For other titles published in the series, go to
www.springer.com/series/3192

The 100 Best Targets for Astrophotography

Ruben Kier

Springer

Ruben Kier
Orange, CT 06477
USA

ISSN 1431-9756

ISBN 978-1-4419-0602-1   e-ISBN 978-1-4419-0603-8
DOI 10.1007/978-1-4419-0603-8

Library of Congress Control Number: 2009928623

© Springer Science+Business Media, LLC 2009
All rights reserved. This work may not be translated or copied in whole or in part without the written permission of the publisher (Springer Science+Business Media, LLC, 233 Spring Street, New York, NY 10013, USA), except for brief excerpts in connection with reviews or scholarly analysis. Use in connection with any form of information storage and retrieval, electronic adaptation, computer software, or by similar or dissimilar methodology now known or hereafter developed is forbidden.
The use in this publication of trade names, trademarks, service marks, and similar terms, even if they are not identified as such, is not to be taken as an expression of opinion as to whether or not they are subject to proprietary rights.

Printed on acid-free paper

Springer is part of Springer Science+Business Media (www.springer.com)

To my parents, Pearl and Ralph, in celebration of their 60th wedding anniversary:

For the nights when you would fall asleep in the car waiting for me at the local observatory, to your support and encouragement of my education, and your enthusiasm about my astrophotography, I am eternally grateful.

To my children, Melanie and Shelley, through whose eyes I have rediscovered the marvels of the cosmos; may you never abandon your sense of wonder at the miracles of nature.

And to my wife, Stephanie, who by her example has motivated me to become a better citizen, physician, teacher, parent, and spouse.

Preface

A picture tells a thousand words. Some of astronomy's best communications and teaching tools are its rich legacy of images. Astroimaging began in 1840 when American astronomer John W. Draper took a 20-minute exposure of the Moon through a 5-inch. Newtonian reflecting telescope. Since then, professional and amateur astronomers, nature photographers, and ordinary people with cameras have created millions of celestial images. Some of these photos have led to important discoveries.

In 1888, a photograph of the Andromeda Galaxy (M31) revealed its spiral structure. In 1919, a picture taken during a total solar eclipse confirmed Einstein's theory that massive objects bend starlight. In 1930, American astronomer Clyde Tombaugh discovered Pluto on a photograph of a starry region in Gemini. And in 2004, the Hubble Space Telescope took a million-second-long exposure of a seemingly empty region of space in the constellation Fornax and revealed thousands of distant galaxies.

Likewise, amateur astronomers have made important contributions. Some of their images have shown previously unknown comets, asteroids, and supernovae. Most amateurs, however, image celestial objects for the sheer joy of it. They produce impressive results using techniques unknown to astronomy only a decade ago.

Like a telescope, astroimaging encompasses two aspects. Most people picture a telescope as a long tube through which you view celestial objects. But the optical tube assembly is just half of a high-quality telescope. Without the accompanying mount, the tube would be of little value.

So it is with astroimaging. If all it took to be an astroimager was to sit for long hours at a telescope, many amateur astronomers would be great at it. But there is another facet to producing excellent images – postprocessing. This includes everything that happens after you acquire the raw data, which, in many instances, does not look all that appealing.

Astroimagers spend years developing and refining their techniques. A high-quality image will show intricate detail, have a wide dynamic range, and exhibit the correct color rendition.

As Photo Editor of the world's best-selling astronomy magazine, I receive thousands of images each year. Of those, we publish perhaps 100. Ruben Kier's images are well represented every year.

By following his instructions, you, too, can produce equally beautiful astroimages. When you do, be sure and send them to me for the magazine. Send your very best work. Remember, you will be competing with Ruben.

Senior Editor/Photo Editor, *Astronomy* magazine   Michael E. Bakich

Acknowledgements

I would like to thank my editors at Springer, Maury Solomon and John Watson, for supporting my interest in producing this book. Their advice and guidance have been invaluable in bringing this project to fruition.

This book is a compilation of many influences. I first became interested in CCD imaging after hearing Robert Gendler speak at the Connecticut Star Party in 2001. Since then, his lectures on luminance layering and hybrid imaging have influenced my processing routines, and the images in his book, *A Year in the Life of the Universe*, have been an example of what I aspire to create with more humble equipment. Ron Wodaski's book, *The New CCD Astronomy*, served for several years as my basic text for CCD techniques. Mike Rice, at the New Mexico Skies Guest Observatory, provided me with invaluable hands-on training. Neil Fleming has helped me to streamline my focusing routines. Scott Ireland's textbook and Jerry Lodriguss' CD on Photoshop techniques have improved my processing skills. Stephen O'Meara's observing guides have helped me understand more about my celestial targets, and his *Hidden Treasures* have helped me select some of the more obscure targets in my book. Robert Burnham's *Celestial Handbook* has provided valuable background detail on many of the Best Targets. Most of the object data was compiled either from Robert Strong's *Sky Atlas 2000 Companion* or from CCDSoft's program *TheSkySix*. Jay Pasachoff and Alex Filippenko's college textbook, *The Cosmos*, has been a resource for clarifying many of the more difficult concepts in astronomy. Ray Gralak has kindly provided data on the periodic error of several mounts. Finally, I thank Michael Bakich, Photo Editor for *Astronomy* magazine, for selecting many of my favorite images for the Reader Gallery over the past several years. Each inclusion reinforces my interest and enjoyment of this exciting hobby.

Contents

Preface vii

Acknowledgements ix

Introduction xvii

Section 1 Best 100 Astrophotography Targets

Chapter 1 January: Mostly Nebulae 3
 January 1: Spiral Galaxy IC 342 4
 January 2: Pleiades Open Cluster 7
 January 6: California Nebula 10
 January 21: Witch Head Nebula 13
 January 24: Flaming Star Nebula 16
 January 26: Tadpole Emission Nebula 19
 January 27: Open Cluster M38 with Open
  Cluster NGC 1907 22
 January 29: Crab Nebula Supernova Remnant ... 25
 January 29: Orion Nebula and Running
  Man Nebula 28
 January 31: Horsehead and Flame Nebulae 31

Chapter 2 February: Clusters and Nebulae 35
 February 1: Reflection Nebula M78 36

February 2: Open Cluster M37	39
February 6: Angel Nebula	42
February 6: Open Clusters M35 and NGC 2158	45
February 9: Jellyfish Nebula	48
February 12: Rosette Nebula and Cluster	51
February 14: Cone Nebula and Christmas Tree Cluster	54
February 24: Thor's Helmet	57
February 27: Medusa Nebula	60
February 27: Eskimo Nebula	63

Chapter 3 March: Clusters and Galaxies ... 67
March 1: Open Clusters M46 and M47	68
March 2: Spiral Galaxy NGC 2403	71
March 20: Ancient Open Cluster M67	74
March 30: Barred Spiral Galaxy NGC 2903	77

Chapter 4 April: Galaxy Pairs and Groups ... 81
April 5: Galaxies M81 and M82	82
April 10: Little Pinwheel Galaxy	87
April 11: Hickson 44 Galaxy Group	90
April 17: Galaxy Pair M95 and M96	93
April 24: Owl Nebula M97 with Galaxy M108	96
April 26: Galaxy Trio in Leo	99
April 26: Hamburger Galaxy NGC 3628	102
April 30: Galaxy Pair NGC 3718 and NGC 3729	105

Chapter 5 May: Diversity of Galaxy Shapes ... 109
May 6: Galaxy M109	110
May 11: Silver Needle Galaxy	113
May 11: Galaxy M106	116
May 12: Galaxies M100 and NGC 4312	119
May 13: Markarian's Chain	122
May 16: Needle Galaxy	125
May 17: Sombrero Galaxy	128
May 17: Whale Galaxy and Hockey Stick Galaxy	131
May 19: Ringed Galaxy NGC 4725	134
May 19: Spiral Galaxy M94	137
May 21: Black Eye Galaxy M64	140
May 24: Sunflower Galaxy M63	142
May 29: Whirlpool Galaxy M51	145

Chapter 6 June: Globular Clusters and More Galaxies ... 149
June 1: Globular Cluster M3	150
June 6: Pinwheel Galaxy M101	153

Contents

June 25: Splinter Galaxy	156
June 26: Globular Cluster M5	159

Chapter 7   **July: Just Globular Clusters** 163
July 17: Great Hercules Cluster 164
July 18: Globular Cluster M12 167

Chapter 8   **August: Planetary and Emission Nebulae** 171
August 5: Cat's Eye Nebula 172
August 6–7: Trifid and Lagoon Nebulae 175
August 7: Lagoon Nebula 178
August 10: Eagle Nebula 181
August 11: Swan Nebula 184
August 15: Globular Cluster M22 187
August 18: Wild Duck Cluster 190
August 19: Ring Nebula 193

Chapter 9   **September: Autumn Assortment** 197
September 1: Barnard's Galaxy 198
September 5: Dumbbell Nebula 201
September 8: Crescent Nebula 204
September 14: Fireworks Galaxy and Cluster
NGC 6939 207
September 17: Veil Nebula, "Witch's Broom"
of Veil 209
September 19: Eastern Loop of Veil:
The Network Nebula 211
September 18: Pelican Nebula 214
September 20: North American Nebula 217
September 20: Fetus Nebula 220
September 21: Iris Nebula 223
September 28 and 29: Globular Clusters M15
and M2 226
September 30: Nebula IC 1396
and the Elephant's Trunk 229

Chapter 10   **October: Halloween Treats** 233
October 4: Cocoon Nebula 234
October 9: Wolf's Cave and the Cepheus Flare 237
October 13: Helix Nebula 240
October 14: Stephan's Quintet 243
October 15: Deer Lick Galaxy Group 246
October 19: Flying Horse Nebula 249
October 20: Cave Nebula 252

October 26: Bubble Nebula and M52 Cluster 255
October 27: Blue Snowball Nebula 258

Chapter 11 November: The Great Galaxies 261
November 1: Andromeda Galaxy and Companions ... 262
November 2: Skull Nebula and Galaxy NGC 255 265
November 17: Sculptor Galaxy 268
November 18: Pacman Nebula 271
November 25: ET Cluster 274
November 29: Triangulum Galaxy 277
November 30: Spiral Galaxy M74 280

Chapter 12 December: Celestial Potpourri 283
December 1: Little Dumbbell Nebula 284
December 5: Nautilus Galaxy NGC 772 287
December 10: Double Cluster 290
December 11: Outer Limits Galaxy 292
December 11: Barred Spiral Galaxy NGC 925 295
December 14: Heart Nebula and December 18:
Soul Nebula 298
December 18: Spiral Galaxy M77 301

Section 2 Getting Started in CCD Imaging

Chapter 13 Equipment for Astrophotography 307
First and Foremost, the Mount 308
Beginner Scopes for Imaging 309
Choosing a Camera 311
Autoguiders 313
Investing Wisely in Software 315

Chapter 14 Acquiring the Image 317
Siting the Telescope 318
Polar Alignment 320
Choosing the Target 322
Calculating Your Fields of View and Scale 324
Finding, Centering, and Framing the Target 326
Focusing 328
Autoguiding 330
Exposing 332
Dark Frames 334
Flat Fields 336

Contents

Chapter 15	**The Order of Image Processing**	339
	Image Reduction/Calibration	340
	Optional Steps	341
	Aligning Images	342
	Combining Images	343
	Deconvolution	344
	Color Combining	345
	Histograms and Curves	346
	Luminance Layering	348
	Color Enrichment	350
	Image Sharpening and Blurring	351
	Dealing out Gradients	353
	Final Cleanup	354

About the Author 355

Index 357

Introduction

How This Book Differs from Observing Lists

Over 200 years ago, many of the celestial treasures on the following pages were cataloged by Charles Messier and William Hershel using telescopes primitive by today's standards. These catalogs have formed the basis of most amateur astronomers' targets for observing. The most famous is Messier's Catalog of 109 objects. Despite their popularity with visual astronomers, Charles Messier's choices were neither the brightest nor the most beautiful through the eyepiece. His list was compiled to define objects that might be confused with comets by other comet hunters – in other words, a list of potential mistakes. Entire regions of the sky, which fell outside of the area where comets might be found, were excluded from his list. This may explain how several bright deep sky objects such as the Double Cluster in Perseus were excluded.

In this century, the growth in quality and accessibility of amateur telescopes has driven an explosion of observing lists. The Herschel 400 list, compiled more than 30 years ago by the Ancient City Astronomy Club of Florida, includes objects selected from Herschel's General Catalog that would "challenge" observers with telescopes 6 in. or larger. In 1995, Patrick Moore published the Caldwell list (his legal last name is

Caldwell-Moore) of 109 objects, which includes both bright and dim objects excluded by Messier. His selection includes some small and challenging targets, not just the crowd pleasers. Most recently, in 2007, Stephen O'Meara published a list of 109 "hidden treasures" that seeks to fill in the gap left by the Messier and Caldwell lists. Like the Caldwell list, O'Meara's list dips deep into the southern hemisphere.

These famous lists are excellent for visual astronomy but can be disappointing for the astrophotographer. For example, a sparse open star cluster sparkles at the eyepiece but can be uninspiring as an image. A small planetary nebula may be striking visually but may be too small to show interesting detail in a photograph. On the other hand, many nebulae that are faint to the eye can have striking texture and hue on long exposures. Spiral galaxies blossom into a rich diversity of shapes and colors.

How the Targets Were Chosen

This book showcases the 100 best targets available to backyard astrophotographers in the Northern Hemisphere. These selections include 48 Messier, 28 Caldwell, and 13 O'Meara objects, plus several others cited in catalogs by Arp, Hickson, Sharpless, and Barnard. Almost a third of the targets can be framed to include multiple objects. The criteria for inclusion were simple:

- Does the image inspire the viewer?

- Is the object bright enough to image with a backyard amateur telescope, an average CCD camera, and 2–3 h of exposure?

- Is it large enough to show detail, usually 5 arcmin or more?

- Can it be photographed successfully from northern latitudes? (This usually requires a declination above −25°.)

  Other features favoring inclusion are:

- Among similar objects, is it the easiest to image because of declination, size, color, or brightness?
- Can the object be framed with a second object to create a more dynamic image?

The images on the following pages represent what an average amateur can expect to accomplish with some practice and effort. In some cases, with

Introduction

modest equipment, you can expect to approach the efforts of the best imagers who use large format cameras through sophisticated astrographs on monolithic mounts. In other cases, a more humble image can be quite satisfying. In every case, you will be capturing views of celestial objects matching or surpassing the photographs taken only decades ago from the finest observatories such as Palomar, Lick, Mt. Wilson, and Lowell.

Our goal is to walk you through the steps of choosing an object, acquiring the image, and then processing the data until you say, "Wow!" The date at the top of each entry indicates the day when the target is on the meridian at 9 p.m. standard time (10 p.m. daylight savings time) for observers in the center of their time zone. Accompanying each illustration, the first paragraph gives a brief background about the object and its significance. A second paragraph advises on how to acquire the image, including suggestions on framing, exposures, binning, guiding, and filters. A third paragraph provides specific suggestions for processing the object.

The second section of this book is a brief introduction to digital astrophotography, with an emphasis on CCD imaging with moderately priced equipment. For beginners, we have tried to keep explanations straightforward. For more advanced imagers, we have included some pointers gained from experience (i.e., mistakes!).

What Are We Looking at?

At the beginning of time, the Big Bang spread mostly hydrogen and some helium into the early inflating universe. These gases condensed into the giant luminous stars that populated young, unstructured galaxies, barren of planets, burning brightly yet briefly, and then dying violently in the cataclysm of supernova explosions. These blasts created the heavy elements, which produced the wide expanses of interstellar dust that drift among the Milky Way's spiral arms. Other gravitational interactions collapsed this dust and gas into new generations of stars. These descendants have evolved into the spectacular diversity of clusters, nebulae, and galaxies that entice the astronomer in all of us.

The following 100 targets include 21 images dominated by emission nebulae, often in combination with an open cluster. Clouds of gas and dust abound in our galaxy. In interstellar space, far from any star, the gas is cold and dark. In some cases, a portion of the gas cloud collapses under its own gravity to generate young bright O and B type stars, which form an associated open star cluster. The ultraviolet light from these new stars ionizes the surrounding gas, stripping off electrons. When the electrons recombine with ions, they release photons of discrete energy (and thus discrete wavelengths) as the electrons cascade back down to a lower energy level. One of these wavelengths is termed the hydrogen alpha band, which contributes to the red glow of emission nebula and allows us to image these nebulae with H-alpha filters.

These pages also include six images dominated by reflection nebulae. In these cases, interstellar dust contributes up to 2% of the mass in a gas cloud. These microscopic particles of dust collectively reflect the blue light of nearby stars, both because they are usually associated with young hot stars that shine blue-white, and because dust is more effective in reflecting blue light than red light. For much the same reason, the dust in Earth's atmosphere reflects sunlight to make our sky blue. However, as the dust becomes denser, or collects farther from any stars, it may actually block the light of stars or the glow of hydrogen emission, giving rise to a dark nebula. Depending on the distribution of gas and dust, many of these images contain mixtures of emission, reflection, and dark nebulosity.

With time, the solar winds of these young stars disperse surrounding clouds of dust and gas. They leave the open star clusters shining alone, as illustrated in nine images. The younger clusters will appear to shine blue-white in our images, not because their stars are all the same color but because the red and yellow stars are much dimmer than the blue and white stars. As star clusters age, the short-lived blue stars expire, and the remaining white and yellow stars begin to dominate our images.

An extreme example of stellar evolution is seen in the seven images of globular clusters. These ancient dense clusters of hundreds of thousands of stars are at least 12 billion years old. White and yellow stars comprise most of the stars in the cluster, appearing as countless tiny points of light that appear to merge in the core of the globular cluster. Scattered red giants, swollen to luminosity a hundred times greater at the end of their lifespan, become the largest and brightest individual stars within the image.

Eighteen images show stars at the end of their stellar evolution, either as planetary nebulae (11 images), Wolf–Rayet stars (3 images), and supernova remnants (4 images). Planetary nebulae are created by dying stars that blow off their outer layers once their central supply of fuel is exhausted. The remnant central star is a dense "white dwarf" that can no longer support nuclear reactions, yet provides the ultraviolet torch that illuminates the outer shells of gas. Whereas the lifespan of the star is measured in billions of years, the stunning colors of the planetary nebula last a mere thousand years. Thus, upon their deaths, these stars impart to the universe a brief but memorable gift of beauty.

Thirty-nine images emphasize galaxies. Most are variations of spiral galaxies, including face-on pinwheels, edge-on cylinders, and several

Introduction

peculiar galaxies. Elliptical galaxies are only shown as companions to spirals. Ellipticals are relatively inert, composed largely of older stars, and lacking sufficient interstellar gas or dust to undergo much new star formation. Therefore, elliptical galaxies also lack the red glowing hydrogen-alpha regions, the blue-white clusters of newly formed stars, and the dark dust lanes that enrich images of spiral galaxies.

Thirteen of these galaxy images show groups of two or more galaxies. We imagine galaxies drifting in the endless sea of space, like isolated universes, stable and timeless. Yet most galaxies are associated into clusters and superclusters, constantly interacting, occasionally disrupting each other, and sometimes even merging, with each appearing unique in shape, size, and structure.

As you image these wonders, remember to contemplate their role in the evolution of our universe.

The excitement of any hobby is based on discovery and creativity. Many routes can be followed to the final image, which merges art with science. Adapt these techniques to your equipment, environment, and inclination, and experiment after taking the advice of others. Allow your images to become a unique interpretation of each cluster, nebula, and galaxy.

SECTION ONE

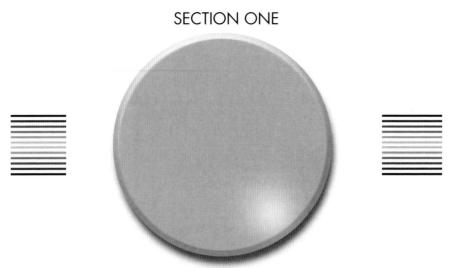

Best 100 Astrophotography Targets

CHAPTER ONE

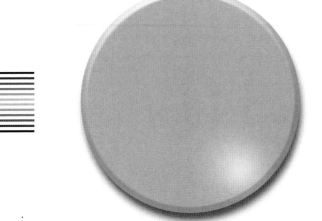

January: Mostly Nebulae

January 1: Spiral Galaxy IC 342

Designation	IC 342
Other names	Caldwell 5
Right ascension	03 h 46.8 min
Declination	+68° 06′
Magnitude	9.2
Size	18 × 17 arcmin
Constellation	Camelopardalis

Spiral Galaxy IC 342 closely resembles our own Milky Way Galaxy. At a distance of between 10 and 14 million light-years, a galaxy of this size would be expected to be one of the brightest in the sky. However, because it lies only 10° above the disk plane of the Milky Way, its light is dimmed tenfold by dust within our own galaxy, and therefore was not discovered until 1895. Several prominent hydrogen clouds, termed H2 regions, populate its spiral arms.

Imaging. Frame this galaxy with a field of view of at least 25 arcmin to define the full extent of its spiral arms, although a field up to 40 arcmin can yield pleasing details. Imaging IC 342 can be a challenge despite its large size and relatively bright magnitude of 9.2. Because it is large, this magnitude is spread out, yielding a low surface brightness. Choose a night of excellent transparency, without the distraction of the Moon. A dark sky is especially helpful. If you must image from suburban skies, consider a light pollution filter such as the IDAS. Gather as much luminance data as possible, perhaps over more than one night. For this image, a high-resolution luminance was obtained with a large telescope and lower resolution color channels with a medium-sized telescope. Single-shot color cameras would require very long exposures to capture IC 342.

Processing. Begin processing your exposures of IC 342 with routine calibration, alignment, and combination of images. After balancing color, gently enhance your color intensity. Do not be surprised if the color remains bland or "muddy" within IC 342. You are imaging through galactic dust, which both scatters the blue light of young star clusters and dulls the red of emission nebulae. Sharpen the brighter regions of the galaxy but smooth its outer arms (Fig. 1.1).

January: Mostly Nebulae

Fig. 1.1. Spiral Galaxy IC 342. East–northeast is up.

Cameras	ST10XME (luminance), ST2000XM (color)
Telescopes	12-in. Meade LX200R at f/7 (luminance)
	5.5-in. TEC refractor at f/7 (color)
Field of view	41 × 31 arcmin
Exposures	Luminance 21 × 5 min, unbinned
	R 6 × 10 min, G 3 × 10 min,
	B 4 × 10 min, unbinned
Scale	1.6 arcsec/pixel
Limiting magnitude	6.0

January 2: Pleiades Open Cluster

Designation	Messier 45
Other names	Pleiades
Right ascension	03 h 47.5 min
Declination	+24° 07'
Magnitude	1.2
Size	110 arcmin
Constellation	Taurus

The M45 Pleiades Cluster contains 500 stars spread across a sphere 14 light-years wide at a distance of 400 light-years. Termed the "Seven Sisters" in mythology, at least seven of the stars can be seen with the naked eye, making a small dipper shape. A telescope can show faint nebulosity of interstellar dust that blossoms in CCD images. The cluster is moving with a radial velocity different from the nebulosity, suggesting that its stars are crossing the path of dust in a molecular cloud (Fig. 1.2).

Imaging. The Pleiades is a challenge because the stars are bright compared to the surrounding nebula. Routine RGB methods or single-shot color is suggested. A separate luminance channel is neither necessary nor desirable. Antiblooming cameras help suppress the excess light of the bright stars from spilling into adjacent pixels in the camera. Even with antiblooming protection, short exposures are required to keep the brighter stars in M45 from overwhelming the image. Consider exposures of 2 min or less if using an antiblooming camera or of 1 min or less if using a non-antiblooming camera. To capture the nebulosity, try to obtain a dozen or more exposures with each filter. Either a short focal length telescope or camera lens that yields a wide field of view is best for framing M45. Because camera lenses and semiapochromatic refractors will focus differently with each filter, make sure to refocus with each filter change.

Processing. When aligning images from different filters, be careful about slight differences in scale. If your focus varied much between filters, the scale will be altered. Some astronomical processing programs will rescale images during alignment, and others will not. If your program does not rescale the images, then wait to combine your color channels in Photoshop and rescale there. The halos around the brighter

Fig. 1.2. Pleiades Star Cluster M45. East is up.

January: Mostly Nebulae

stars are unavoidable, and are caused by internal reflections between the camera and the overlying filters. These can be toned down in Photoshop and other programs by selecting the halos and dimming them (see section "Final Cleanup" in Chap. 15).

Camera	ST2000XM
Telescope	3.5-in. Takahashi refractor at f/4.5
Field of view	100×75 arcmin
Exposures	R and G each 25×2 min, B 45×2 min, unbinned
Scale	3.8 arcsec/pixel
Limiting magnitude	6.0

January 6: California Nebula

Designation	NGC 1499
Other names	Sharpless 2-220
Right ascension	04 h 00.7 min
Declination	+36° 37′
Magnitude	–
Size	145 × 40 arcmin
Constellation	Perseus

The California Nebula derives its name from its characteristic shape on long exposure photographs. This gas cloud is located 2,000 light-years away with a total mass of 240 suns and is illuminated by the bright star to its right in this photo. Visually, this emission nebula is hard to observe even with large telescopes, because its dim light is spread over an area four times the size of the Moon (Fig. 1.3).

Imaging. The California Nebula requires a large field of view. Unless you have a giant imaging chip, or are willing to devote the effort to create a mosaic image, your best bet may be to use a camera lens. An old manual 300-mm ED lens purchased on EBay for a fraction of the cost of a new lens was used here. If you use a camera lens, remember to refocus between filters. This image was acquired in the light-polluted skies of Connecticut, using a Hydrogen-alpha narrow band filter for the luminance. Narrow band filters allow high-quality imaging from suburban areas by excluding most light pollution. If you do not have an H-alpha filter, you can still get excellent results with a red filter for luminance. You can bin the images to gather light faster, but you can acquire higher resolution images unbinned. Tracking is not very demanding with the short focal length.

Processing. Because the California Nebula is a purely emission nebula, you would not lose any detail by keeping the H-alpha or red exposures for luminance. A HaRGB- or RRGB-layered luminance works well. A pure H-alpha luminance may create dark halos around small stars. To prevent this, you can match the star sizes of H-alpha luminance to the RGB channels either by blending the H-alpha image with the red channel, using pixel math to combine red with H-alpha, or using the lighten or screen mode in Photoshop to superimpose red on H-alpha.

January: Mostly Nebulae

Fig. 1.3. California Nebula NGC 1499. West is up.

Camera	ST10XME
Telescope	300-mm Nikon camera lens at *f*/4.5 lens
Field of view	167 × 112 arcmin
Exposures	Luminance H-alpha, 24 × 5 binned 2 × 2
	R 3 × 5 min, G 4 × 5 min,
	B 3 × 7 min, binned 2 × 2
Scale	9.2 arcsec/pixel
Limiting magnitude	3.5

January 21: Witch Head Nebula

Designation	IC 2118
Other names	Witch Head
Right ascension	05 h 06.9 min
Declination	–07° 13′
Magnitude	–
Size	180 × 60 arcmin
Constellation	Eridanus

The faint Witch Head Nebula, 1,000 light-years distant, is composed of small dust grains reflecting blue light from the nearby brilliant star Rigel, which is just beyond the field of this image on the right. Can you make out the large chin, round open mouth, and pointed nose of the wicked witch? Like children lying back on the grass on a summer day, astronomers also gaze at the sky, imagining shapes in ethereal wisps of dust and gas. Human nature tries to create order from chaos (Fig. 1.4).

Imaging. The Witch Head requires a large field of view. This image spans an area $2 \times 1.5°$, which barely covers the object. You may prefer to use a camera lens with a shorter focal length to better frame the object. Any light pollution will obscure the nebula. Furthermore, light pollution will introduce background gradients that are exaggerated by the large field of view and (from northern latitudes) by the low altitude of the object. Minimize these effects by trying to image close to the meridian when the Witch Head is highest. A single-shot color camera will require multiple long exposures under a dark sky to capture the dim nebulosity. Tracking is not very demanding with the large field of view.

Processing. After routine calibration, alignment, and combination of exposures, use either digital development (DDP) or multiple applications of curves/levels to bring out the dim nebulosity without bloating the stars. Gently sharpen just the nebulosity with a selection or a layer mask to enhance the detail; avoid sharpening in dim areas and in the background to prevent annoying noise. Finally, after carefully balancing the color, boost the color saturation to display the rich blue and violet nebulosity. Very long exposures from a dark sky may reveal faint red and blue nebulosity in the background, so be careful when correcting background gradients.

Fig. 1.4. Witch Head Nebula in Eridanus. South is up.

January: Mostly Nebulae

Camera	ST10XME
Telescope	3.5-in. Takahashi refractor at $f/4.5$
Field of view	126 × 85 arcmin
Exposures	Luminance clear filter 15 × 5 min, unbinned R, G, and B filters each 6 × 5 min, binned 2 × 2
Scale	3.5 arcsec/pixel
Limiting magnitude	6.0

January 24: Flaming Star Nebula

Designation	IC 405
Other names	Caldwell 31
Right ascension	05 h 16.2 min
Declination	+34° 16'
Magnitude	–
Size	30 × 19 arcmin
Constellation	Auriga

The Flaming Star Nebula is part of a molecular cloud illuminated by the "runaway" star AE Aurigae. This bright star is a transient visitor to this region, ejected from the Orion Nebula by the collision of two binary star groups. Ultraviolet radiation from the star ionizes and excites hydrogen gas glows to glow red. A smaller region closer to the star shines blue, due to the dust reflecting the starlight. The dimensions quoted for the Flaming Star Nebula exclude a long dimmer tail of nebulosity with a sharp border swept clean by the passage of the runaway star (Fig. 1.5).

Imaging. You can choose to frame just the main brighter region of the Flaming Star Nebula with a field of view of about 40 arcmin, or the larger region as in my second image with a field of 100 arcmin. You can even choose a field of 150 arcmin to include IC 410 (several pages forward) in your frame. The brighter portions of the nebula can be captured with either RGB methods or single-shot color. The dimmer tail in the larger field is revealed with H-alpha exposures.

Processing. If you are using an H-alpha luminance, consider blending this channel with your red channel to create a new blended luminance whose stars will better match your RGB exposures. Enhance the blue portions of the nebulosity with either increased saturation or Photoshop's match color adjustment.

January: Mostly Nebulae

Fig. 1.5. (a) Flaming Star Nebula. West is up. (b) Flaming Star Nebula with tail of nebulosity. Northwest is up.

Camera (close-up)	ST2000XM
Telescopes	5.5-in. TEC refractor at *f*/7
Field of view	41 × 31 arcmin
Exposures	R, G, and B each 3 × 10 min, unbinned
Scale	1.6 arcsec/pixel
Limiting magnitude	6.0

Camera (wide field)	ST10XME
Telescope	3.5-in. Takahashi refractor at *f*/4.5
Field of view	126 × 85 arcmin
Exposures	H-alpha 10 × 5 min, binned 2 × 2
	R and G, each 5 × 5 min, B 11 × 5 min, binned 2 × 2
Scale	7 arcsec/pixel
Limiting magnitude	3.5

January: Mostly Nebulae

January 26: Tadpole Emission Nebula

Designation	IC 410 (nebula)
Other names	NGC 1893 (cluster)
Right ascension	05 h 22.6 min
Declination	+33° 31'
Magnitude	7.8
Size	40 × 30 arcmin
Constellation	Auriga

Early astronomers often detected young open clusters without recognizing the dim gas cloud from which they emerged. The open cluster in the center of the nebula, NGC 1893, was described long before the surrounding emission nebula IC 410. The stellar wind from these young stars sculpts some of nebula's gas and dust into two "tadpoles" at the upper right of the gas cloud (Fig. 1.6).

Imaging. The central star cluster covers a diameter of 11 arcmin, while the nebula extends to 30 arcmin or more. You can either frame IC 410 with a field of view of 50 arcmin, or include the Flaming Star Nebula with a field of 3 × 2°. Although bright enough for single-shot color or RGB methods, more nebulosity will be revealed with an H-alpha or red luminance layer.

Processing. After combining your channels, adjust your histogram with either digital development or curves/levels to bring out the dim nebulosity. A rich luminance in the brighter areas of the nebula can tolerate some aggressive sharpening, especially around the tadpoles. Blend in H-alpha luminance gradually to avoid creating strange halos around brighter stars in the nebula and to prevent the nebulosity from acquiring an unnatural salmon color (see section "Luminance Layering" in Chap. 15).

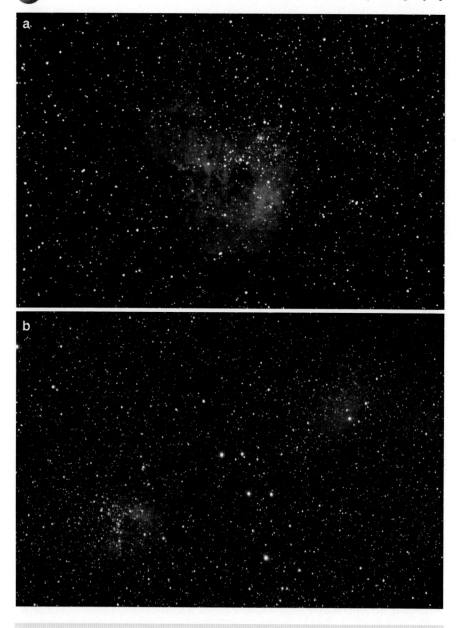

Fig. 1.6. (a) IC 410, the Tadpole Nebula. East is up. (b) The Tadpole Nebula (*left*) and the Flaming Star Nebula (*right*). North is up.

January: Mostly Nebulae

Camera	ST10XME (both images)
Telescope	4-in. Astro-Physics refractor at f/6 (IC 410 only)
Field of view	82 × 55 arcmin
Exposures	H-alpha 15 × 5 min, binned 2 × 2 R 12 × 5 min, G 6 × 5 min, B 11 × 5 min, binned 2 × 2
Scale	4.5 arcsec/pixel
Limiting magnitude	3.5
Telescope	300-mm Nikon camera lens at f/4.5 lens (wide view)
Field of view	167 × 112 arcmin
Exposures	Luminance H-alpha, 15 × 10, unbinned R 13 × 5 mim, G 14 × 5 min, B 25 × 5 min, unbinned
Scale	4.6 arcsec/pixel
Limiting magnitude	6.0

January 27: Open Cluster M38 with Open Cluster NGC 1907

Designation	Messier 38	NGC 1907
Other names	NGC 1912	Melotte 35
Right ascension	05 h 28.7 min	05 h 28.1 min
Declination	+35° 50'	+35° 20'
Magnitude	6.4	8.2
Size	15 × 15 arcmin	6 × 6 arcmin
Constellation	Auriga	Auriga

Open Cluster M38, in the lower left of this view, is one of the three famous Messier clusters in Auriga (along with M36 and M37). Some describe its 100 stars as forming the shape of the Greek letter Pi. Lying at a distance of about 4,000 light-years, this 200 million-year-old cluster is still dominated by young blue-white stars, with a few scattered red giants. Generally, blue-white stars are large and bright, and so burn out earlier than other dimmer yellow stars. Open Cluster NGC1907, to the upper right of the image, is slightly farther away than M38 at a distance of 4,500 light-years. Its 30 stars appear more tightly clustered at least in part because of this greater distance. With an age of 440 million years, fewer blue-white stars survive than in M38 (Fig. 1.7).

Imaging. M38 is much more interesting when framed along with NGC 1907. The two clusters are oriented north–south, but the image is more pleasing with the clusters at an angle. The bright star next to NGC 1907 is a sixth magnitude K2 star, accounting for its orange color, and provides a third point of interest. If you have a non-antiblooming CCD camera, keep your exposures short to keep this bright star from blooming excessively. As with most bright open clusters, routine RGB gives better results than LRGB technique. Single-shot color cameras work well with bright open clusters like M38.

Processing. Like most open clusters, processing is relatively straightforward. Unlike processing routines for nebulae and galaxies, you can allow curves and levels to brighten stars. More stars give a richer image. Many sharpening routines will enhance dimmer stars. Try to avoid moving the white point too far into the histogram, which may rob color from

January: Mostly Nebulae

Fig. 1.7. Open Cluster M38 (*lower left*) with Open Cluster NGC 1907 (*upper right*). Southwest is up.

the stars. After you bring your stars to a desirable brightness, correct for any gradients in your image. For this image that was obtained in a light-polluted suburb, I used GradientXTerminater.

Camera	ST10XME
Telescope	4-in. Astro-Physics refractor at f/6
Field of view	82 × 55 arcmin
Exposures	R 19 × 2 min, G 18 × 2 min, B 18 × 2 min, unbinned
Scale	2.2 arcsec/pixel
Limiting magnitude	3.5

January 29: Crab Nebula Supernova Remnant

Designation	Messier 1
Other names	NGC 1952, Crab nebula
Right ascension	05 h 34.5 min
Declination	+22° 01'
Magnitude	8.0
Size	6 × 4 arcmin
Constellation	Taurus

The Crab Nebula was the first entry in Charles Messier's famous catalog of celestial treasures, and thus designated as M1. Messier was a French astronomer and comet hunter in the 1700s who recorded over 100 "fuzzy objects" which could be mistaken for comets. Little did he suspect that he would be remembered more for his catalog of "mistakes" than for any comet he discovered. M1 is also known as the Crab Nebula, due to its vague resemblance to a horseshoe crab. M1 is a remnant of a supernova, from a star that exploded in 1054 C.E., and chronicled by Chinese astronomers as bright enough to be seen in daylight for days! Type II supernovae occur when a giant star (at least nine times as massive as our sun) runs out of nuclear fuel, begins to collapse, and then rebounding in a huge explosion that can be brighter than an entire galaxy! The filaments in the Crab Nebula are remnants of the original star's atmosphere, energized by synchrotron radiation from the rapidly spinning neutron star (a pulsar) at its core (Fig. 1.8).

Imaging. The Crab Nebula is a small object that benefits from a high-resolution imaging. Wait for a night with steady skies and excellent seeing. An H-alpha luminance can enrich the view of the detailed tendrils, but if the Crab is a small portion of the image, the star sizes or color may be suppressed to create an unnatural appearance. Most images will have background stars as a major portion of the image, so a red luminance or blended red and clear luminance may be better to more gently enrich the nebulosity.

Processing. Small objects require strong detail to carry the entire image. Use your sharpening tools to develop visual interest. Make sure that colors are balanced properly, and that your background is smooth with good star colors.

Fig. 1.8. Crab Nebula. North is up.

January: Mostly Nebulae

Camera	ST10XME
Telescope	12-in. Meade LX200R at f/7
Field of view	23 × 15 arcmin
Exposures	Luminance clear 6 × 5 min, R 6 × 5 min, unbinned
	R and G each 5 × 5 min, B 7 × 5, binned 2 × 2
Scale	0.6 arcsec/pixel
Limiting magnitude	6.0

January 29: Orion Nebula and Running Man Nebula

Designations	NGC 1976	NGC 1977
Other names	Messier 42	Hidden Treasure 32
Right ascension	05 h 35.3 min	05 h 35.3 min
Declination	−5° 23′	−4° 49′
Magnitude	4.5	–
Size	66 × 60 arcmin	20 × 15 arcmin
Constellation	Orion	Orion

The Great Orion Nebula is the archetypal stellar nursery, with stars emerging from clouds of hydrogen gas and dust. The upper object has been called the "Running Man Nebula." Dark intersecting lanes of dust create the apparition of a man's arms spread wide as he runs toward our left. The blue color represents a reflection nebula, with clouds of dust merely reflecting the light of nearby stars (Fig. 1.9).

Imaging. The Orion Nebula appears fabulous framed either alone or together with the Running Man Nebula. The central part of the Orion Nebula, in the region of the trapezium, is easily overexposed. When you begin imaging, do some test exposures of 1, 2, and 3 min to get an idea of the best duration of exposure. Try to show as much detail as possible in the outer areas of the nebula without saturating the core. If you have a non-antiblooming camera, avoid blooming in both the core and in the bright third magnitude star Iota Orion. Routine RGB imagine or a single-shot CCD camera can yield excellent results.

Processing. The biggest challenge in processing the Orion Nebula is in the tremendous differences in brightness between the core around the trapezium and the dim outer arms. Routine histogram stretching either with digital development or with levels and curves is not sufficient. One solution is to process your image twice, once for the central area (core image), and another for the rest of the nebula (nebula image). In Photoshop, you can paste the nebula image over the core image. Select the nebula image, then lasso the overexposed region of the core, feather several pixels, and then cut (edit, cut). Your properly exposed core should shine through. If the transition seems too harsh, step back to the lasso or feathering stages and vary both your selection and feathering. You can also choose to combine the core and nebula images with layer masks.

January: Mostly Nebulae

Fig. 1.9. Orion Nebula and Running Man Nebula. North is up.

Camera	ST2000XM
Telescope	3.5-in. Takahashi refractor at f/4.5
Field of view	100×75 arcmin
Exposures	R 25×2 min, G 20×2 min, B 22×2 min
Scale	3.8 arcsec/pixel
Limiting magnitude	6.0

January 31: Horsehead and Flame Nebulae

Designations	IC 434	NGC 2024
Other names	Barnard 33	Hidden Treasure 34
Right ascension	05 h 40.9 min	05 h 41.9 min
Declination	–2° 27′	–1° 51′
Magnitude	4.5	7.2
Size	60 × 40 arcmin	30 × 30 arcmin
Constellation	Orion	Orion

This wide field image of the Horsehead Nebula includes a total of four nebulae at a distance of 1,500 light-years, all related to the large stellar nursery in Orion that includes the Great Nebula M42. The largest nebula in this group is the deep red emission nebula IC 434, appearing as a waterfall of ionized hydrogen in the lower half of the image. An intervening cloud of interstellar dust creates the silhouette of a horse's head, giving rise to the name of the dark nebula B33. The Flame Nebula at the top, also known as NGC 2024, is another region of glowing hydrogen gas in the shape of a burning bush. Between the Flame and the Horsehead is the smaller blue reflection nebula NGC 2023, caused by a cloud of fine dust reflecting the light of a central star. The brilliant blue star in the upper right of this image, Alnitak, is better known as the left-side star in the belt of Orion, the Hunter (Fig. 1.10).

Imaging. This image is a challenge because of the brilliant second magnitude star Alnitak, compared to the faint H-II regions. A blended luminance using H-alpha exposures enriches the H-II areas while maintaining color balance. The H-alpha exposures can be binned 2 × 2 to speed the imaging time. The IDAS filter for the luminance helps to suppress light pollution gradients, which would be a nightmare amid the complex nebulae. Short exposures reduce blooming from Alnitak and other bright stars. A wider field might frame the objects even nicer.

Processing. Preparing the blended luminance requires some experimentation (see section "Luminance Layering" in Chap. 15). This image blends the IDAS exposures with the H-alpha for the luminance. The red channel used for the RGB color combine is actually a blend of 40% H-alpha with 60% red, which enriches the red color.

Fig. 1.10. Horsehead and Flame Nebulae. North is up.

January: Mostly Nebulae

Camera	ST10XME
Telescope	4-in. Astro-Physics refractor at f/6
Field of view	82 × 55 arcmin
Exposures	Luminance IDAS 34 × 2 min unbinnned
	H-alpha 12 × 2 min, binned 2 × 2
	R 9 × 2 min, G 9 × 2 min, B 15 × 2 min, binned 2 × 2
Scale	2.3 arcsec/pixel
Limiting magnitude	3.5

CHAPTER TWO

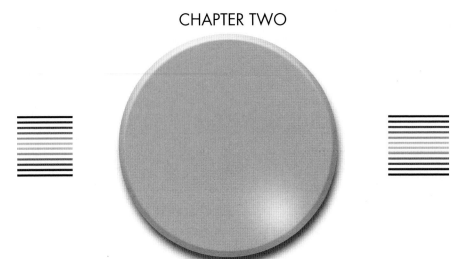

February: Clusters and Nebulae

February 1: Reflection Nebula M78

Designation	NGC 2068	NGC 2071
Other names	Messier 78	
Right ascension	05 h 46.7 min	05 h 47.1 min
Declination	+00° 04'	+00° 18'
Magnitude	8.0	–
Size	8 × 6 arcmin	7 × 5 arcmin
Constellation	Orion	Orion

The Nebulae M78 (bottom) and NGC 2071 (top) are two concentrations of interstellar dust, 1,600 light-years away in Orion. These "reflection" nebulae dimly reflect the blue light of nearby young stars. The other various colors indicate some components of hydrogen and other gases that are excited to higher energy states, which then glow dimly. These nebulae are part of the vast Orion Complex of dust and gas, 1,600 light-years away, which is centered on the Orion Nebula.

Imaging. This can be a frustrating object to image. The areas of reflection nebulosity are ill-defined, and fade gradually to darkness. Even stars appear muted, with many obscured by dark dust. A slightly larger field than shown here might frame the objects better. Because the borders are faint and the object is diffuse, binning all channels including luminance is an option, which I chose for this image. Single-shot color and RGB imaging is possible with long exposures, and has the advantage of yielding more natural colors than LRGB methods.

Processing. Begin processing with routine techniques. After histogram adjustments with DDP or curves/levels, processing becomes more challenging. These reflection nebulae have low intrinsic contrast, so avoid aggressive sharpening that may generate excessive noise. After correcting color balance, enhance color by either boosting saturation or using match color. If your luminance channel causes the color to pale, apply the luminance with less than 100% opacity, flatten the image, and then enhance color again. You can try applying the luminance a second or third time with greater opacity each time to introduce more detail (Fig. 2.1).

February: Clusters and Nebulae

Fig. 2.1. Reflection Nebula M78 (*bottom*) and NGC 2071 (*top*). North is up.

Camera	ST10XME
Telescope	12-in. Meade LX200R at f/7
Field of view	22 × 15 arcmin
Exposures	Luminance clear 12 × 5 min, binned 2 × 2 R and B each 6 × 5 min, G 5 × 5 min, binned 2 × 2
Scale	1.3 arcsec/pixel
Limiting magnitude	6.0

February 2: Open Cluster M37

Designations	NGC 2099
Other names	Messier 37
Right ascension	05 h 53.0 min
Declination	+32° 33′
Magnitude	6.0
Size	15 × 15 arcmin
Constellation	Auriga

M37 is the most beautiful of three open clusters in the constellation Auriga. M37 contains 1,890 suns spread across a region of 20 light-years, at a distance from us of 4,400 light-years. The brilliant orange star in the center is the brightest of about a dozen "red giants" in the cluster, whose visual orange-red color contrasts clearly with surrounding young blue-white stars when viewed through my 5.5-in. refractor. Red giants develop late in the evolution of a star, when nuclear fuel is declining, and the outer envelope of the star expands to hundreds of times its original size. When our sun enters its red giant stage, the edge of the sun will extend to earth's orbit, burning out all life.

Imaging. The open cluster M37 appears nicely framed in fields of 30 arc-min or more. Single-shot color or routine RGB techniques are ideal. Luminance layering has no advantage for this bright object. If you are using RGB filters through a refractor with less than perfect color correction, make sure to refocus with each filter change. If you use a single-shot color camera, keep all of your images above an altitude of 30° to prevent atmospheric refraction form blurring your stars. Open clusters are ideal targets from areas of severe light pollution, because light pollution gradients are easier to subtract.

Processing. For RGB imaging of star clusters, alignment of color channels is critical to avoid asymmetric halos around stars. After combining images, apply histogram adjustments to allow brighter stars to become larger than dimmer stars. Stars are the focus of cluster images, so bright stars should be allowed to become larger than might be tolerated for images of galaxies or nebulae. After correcting color balance, enhance color to distinguish red giant stars from blue-white stars. If artificial sharp halos develop, select your brightest stars and apply a mild Gaussian blur of less than 1 pixel (Fig. 2.2).

Fig. 2.2. Open Cluster M37. Northeast is up.

February: Clusters and Nebulae

Camera	ST10XME
Telescope	5.5-in. TEC refractor at $f/7$
Field of view	52 × 35 arcmin
Exposures	R and G each 6 × 5 min, B 9 × 5 min, unbinned
Scale	1.4 arcsec/pixel
Limiting magnitude	3.5

February 6: Angel Nebula

Designation	NGC 2170
Other names	Van den Bergh 67, 68, and 69
Right ascension	06 h 07.5 min
Declination	–06° 24′
Magnitude	–
Size	10 × 20 arcmin for group
Constellation	Monoceros

Reflection nebula NGC 2170, in the upper left of this image, is the brightest of several blue reflection and red emission nebulae in this region of the constellation Monoceros. The nebulae evoke the appearance of a celestial angel, with the head on the upper right, the blue flowing robes at the bottom, and a red emission nebula at the heart. All of these nebulae are closely associated at a distance of 2,400 light-years, illuminated by young hot B-type stars that are emerging from the giant molecular cloud Mon R2. The NGC 2170 complex is part of a much larger association of hydrogen and dust, stretching from the constellations Monoceros to Orion.

Imaging. This object can be framed either as shown or with a much larger frame of almost triple the size to show additional areas of nebulosity to the right of this image. Faint reflection nebulae require a dark sky site for imaging. You will be disappointed in light-polluted suburbs. Luminance layering with LRGB techniques help to collect light faster for this object. A single-shot color camera would need very long exposures.

Processing. Routine luminance layering techniques are suggested. After histogram adjustments with digital development or levels/curves, gently sharpen the areas of brighter nebulosity. The detail is large scale in the nebulae, so use a radius of 10 pixels or more for your sharpening tools. Avoid including stars in the sharpening areas, which might create annoying dark halos around the stars. After balancing color in your RGB channels, enhance the color by either increasing saturation or using Photoshop's "match color" adjustments. If your color becomes pale when applying the clear luminance channel, first apply it with less than 100% opacity, flatten the image, and then enhance the color again (Fig. 2.3).

February: Clusters and Nebulae

Fig. 2.3. Angel Nebula NGC 2170. Southwest is up.

Camera	ST10XME
Telescope	12-in. Meade LX200R at *f*/7
Field of view	23 × 15 arcmin
Exposures	Luminance clear 22 × 5 min, unbinned R 9 × 5 min, G 8 × 5 min, B 12 × 5 min, binned 2 × 2
Scale	0.6 arcsec/pixel
Limiting magnitude	6.0

February 6: Open Clusters M35 and NGC 2158

Designation	NGC 2168	NGC 2158
Other names	Messier 35	Melotte 40
Right ascension	06 h 08.9 min	06 h 07.4 min
Declination	+24° 20'	+24° 6'
Magnitude	5.1	8.6
Size	25 × 25 arcmin	5 × 5 arcmin
Constellation	Gemini	Gemini

Star Cluster M35 is glittering jewel at the foot of the constellation Gemini, dazzling its viewer through binoculars or a small telescope. Above M35, another cluster lurks dim and distant. Four times farther away than M35, star cluster NGC 2158 is also much older yet richer. The youthful M35, only 100 million years old, shines with bright energetic blue-white suns at a distance of 2,800 light-years. The more mature NGC2158, a billion years old, is illuminated by more gently glowing and long-lasting yellow stars at a distance of 16,000 light-years. M35 also contains yellow stars, but the type O and B blue-white stars dominate our view because they shine much brighter than type G (yellow), K (orange), or M (red) stars (Fig. 2.4).

Imaging. Images of open clusters are more interesting when a second object is in the same view. M35 could be framed by a 40-arcmin field, but a larger field of 70 × 50 arcmin allows inclusion of the smaller cluster NGC 2158. As with other open clusters, either RGB techniques or single-shot color cameras are suggested. Luminance layering is not necessary. RGB imagers using a refractor with less than perfect color correction should refocus with every filter change. Single-shot color imagers should try to avoid imaging below 30° altitude to prevent blurring by atmospheric refraction. Keep exposures short enough to suppress blooming spikes.

Processing. As with all open clusters, carefully align color channels to prevent ugly asymmetric color halos. See the discussion for processing open cluster M37 (section "February 2: Open Cluster M37") for additional suggestions.

Fig. 2.4. Open Clusters M35 and NGC 2158. West is up.

February: Clusters and Nebulae

Camera	ST10XME
Telescope	4-in. Astro-Physics refractor at f/6
Field of view	82 × 55 arcmin
Exposures	Luminance IDAS 21 × 2 min, unbinned R 6 × 2 min, G 5 × 2 min, B 10 × 2 min, binned 2 × 2
Scale	2.3 arcsec/pixel
Limiting magnitude	3.5

February 9: Jellyfish Nebula

Designations	IC 443
Other names	
Right ascension	06 h 16.9 min
Declination	+22° 47'
Magnitude	–
Size	50 × 40 arcmin
Constellation	Gemini

The Jellyfish Nebula, IC 443, is the remnant of a supernova explosion 8,000 years ago in Gemini. Energetic shock waves from the supernova excite hydrogen within a molecular cloud, creating lattice-like filaments of luminosity.

Imaging. You have several choices for framing the Jellyfish Nebula. A large field of view as in this image allows inclusion of part of a nearby hydrogen cloud, which shines a pale red at the top of the image. An even larger field would show more of the pale nebulosity. A smaller field of 80 × 60 arcmin would frame the Jellyfish Nebula more tightly, and allow more detail to be depicted. Orient your image so that blooming spikes from the red giant star Propus of magnitude 3.3 do not extend through the heart of the nebula. Because most of the image is red nebulosity, luminance layering with a filtered luminance is ideal. From the suburbs, an H-alpha luminance helps to suppress the impact of light pollution. From a dark sky site, a red luminance could be used (see section "Luminance Layering" in Chap. 15). If you use a single-shot color camera, this dim object would require very long exposures.

Processing. After calibrating your images, correct blooming spikes around Propus (at the bottom of this image) before aligning the images. After aligning and combining the images, adjust histograms routinely with either digital development or curves/levels. If you use an H-alpha luminance, consider blending it into your red channel to give a better match to your H-alpha luminance. Similarly, you may wish to blend the red channel into your H-alpha luminance to provide better matching of star sizes. When applying an H-alpha luminance in Photoshop, consider using an opacity of under 100%. Finish by correcting any unnatural bloating around the bright star Propus (Fig. 2.5).

February: Clusters and Nebulae

Fig. 2.5. Jellyfish Nebula IC 443. East is up.

Camera	ST10XME
Telescope	3.5-in. Takahashi refractor at $f/4.5$
Field of view	126 × 85 arcmin
Exposures	Luminance Ha filter 6 × 10 min, unbinned R and G each 4 × 5 min, B 6 × 5 min, unbinned
Scale	3.5 arcsec/pixel
Limiting magnitude	6.0

February 12: Rosette Nebula and Cluster

Designation	NGC 2237-9	NGC 2244
Other names	Caldwell 49, Nebula	Caldwell 50, Cluster
Right ascension	06 h 32.3 min	06 h 32.4 min
Declination	+05° 03'	+04° 52'
Magnitude	–	4.8
Size	80 × 60 arcmin	24 arcmin
Constellation	Monoceros	Monoceros

The Rosette is one of the largest nebulae in our galaxy, measuring 115 light-years across. This cloud of gas and dust is intrinsically three times as large as the Orion Nebula, but three times farther away at a distance of 4,900 light-years. A young star cluster NGC 2244 is emerging from the Rosette Nebula, with 100 stars spread across a half degree at the heart of the nebula. Collectively, the stars in the cluster are visible to the naked eye. Ultraviolet light from these central suns ionizes and excites the surrounding gas from which these stars were born. Darker regions of dust, best seen in the upper rim of the nebula, represent Bok Globules. Bok Globules are denser clouds of dust and gas that may be precursors of new solar systems.

Imaging. To fully encompass the Rosette Nebula, a large field of view is needed. My image includes about 1.5 × 2°, which frames the nebula nicely. The nebula is bright enough to be imaged with single-shot color or RGB methods, but the richest nebulosity will come from a filtered luminance. From a dark sky site, a red luminance works best, as in my image. From a light-polluted suburb, an H-alpha luminance can help suppress sky background. Do not forget to refocus between filter changes if your telescope has less than perfect color correction.

Processing. Process luminance layering as described in the section "February 9: Jellyfish Nebula". Apply some sharpening to the Bok Globule regions. Adjust your color balance to allow the central bright stars to shine bluish-white. Be aware that the center of the Rosette Nebula has a reflection component, and should shine more magenta than pure red, with a transition to pure red farther out (Fig. 2.6).

Fig. 2.6. Rosette Nebula. Northwest is up.

February: Clusters and Nebulae

Camera	ST10XME
Telescope	3.5-in. Takahashi refractor at f/4.5
Field of view	126 × 85 arcmin
Exposures	Luminance used red channel exposure R 20×3 min, G 14×3 min, B 13×3 min, unbinned
Scale	3.5 arcsec/pixel
Limiting magnitude	6.0

February 14: Cone Nebula and Christmas Tree Cluster

Designation	NGC 2264
Other names	Hidden Treasure 38
Right ascension	06 h 41.1 min
Declination	+09° 53'
Magnitude	3.9
Size	20 arcmin
Constellation	Monoceros

The Cone Nebula is just a small dark fragment of a vast star-forming region, 2,600 light-years distant. The blood-red glow arises from clouds of hydrogen gas that are excited by the ultraviolet light of nearby young type O and B stars. Above the Cone Nebula, the loose triangular cluster of these stars is called the Christmas Tree cluster. At the top of the second image, interstellar dust reflects the light of the brightest stars in the cluster, glowing blue, at the edge of the Foxfur Nebula.

Imaging. The region of nebulosity around the Cone Nebula is vast. You can frame either a large region up to $3 \times 2°$ with a camera lens or focus in on the cone with a smaller field. From the suburbs, an H-alpha luminance reveals nice detail but light pollution will drown out the faint blue reflection nebula.

Processing. See suggestions for the Jellyfish and Rosette in the sections "February 9: Jellyfish Nebula" and "February 12: Rosette Nebula and Cluster" (Fig. 2.7).

February: Clusters and Nebulae

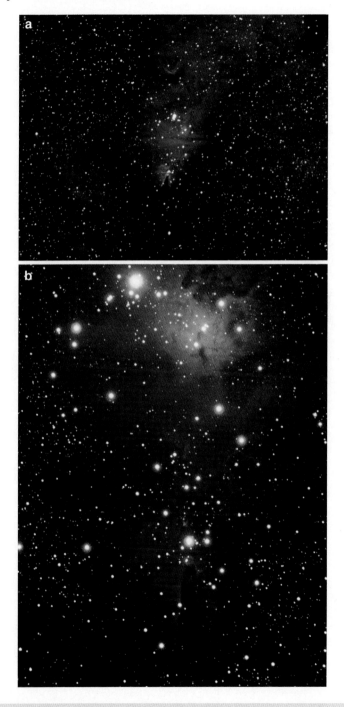

Fig. 2.7. (**a**) Cone Nebula, wide field. North is up. (**b**) Cone Nebula. North is up.

Camera (wide-field)	ST10XME
Telescope	300-mm Nikon camera lens at $f/4.5$
Field of view	167×112 cropped to 147×111 arcmin
Exposures	Ha 6×10 min, G 5×3 min, B 5×3 min, binned 2×2
Scale	9.2 arcsec/pixel
Limiting magnitude	3.5

Camera	ST2000XM
Telescope	5.5-in. TEC refractor at $f/7$
Field of view	41×31 arcmin
Exposures	R 12×10 min, G 6×10 min, B 8×10 min, unbinned
Scale	1.6 arcsec/pixel
Limiting magnitude	6.0

February 24: Thor's Helmet

Designation	NGC 2359
Other names	
Right ascension	07 h 18.6 min
Declination	−13° 12′
Magnitude	–
Size	16 × 8 arcmin
Constellation	Canis Major

Fifteen thousand light-years away in the direction of constellation Canis Major, Thor's Helmet is a complex nebula illuminated by a bright rare Wolf–Rayet type star. High-velocity stellar winds from this energetic star compress, excite, and illuminate surrounding gas and dust to create a bubble of illumination, similar to the process seen in the Bubble Nebula.

Imaging. Thor's Helmet is among the most difficult of the 100 Best Targets for imaging. Its small size requires at least a medium focal length for proper image scale. A field of 30 × 20 arcmin is ideal, but larger frames as in my image are acceptable. Thor's Helmet is also dim and masked by light pollution. From northern latitudes, luminance layering with a clear filter may become blurred by atmospheric diffraction. Its low declination also contributes to atmospheric extinction, which can mute blue colors around the central bubble. Try to obtain your blue exposures and clear luminance exposures near the meridian when its altitude is highest. H-alpha exposures can help detect the dim outer arms of Thor's Helmet, but can overwhelm the weaker blue and green signal of the central region.

Processing. Most processing steps are routine. The object is faint, so signal is usually weak in most of the nebula. Limit sharpening to the central areas with the strongest signal. If you blend H-alpha exposures into your red channel, the red nebulosity may overpower the other channels and obscure the teal central colors. Therefore, if you have obtained an H-alpha luminance, only apply it with routine LRGB methods at less than 50% opacity. Do not force your color balance to the point of unnatural star colors. Finish your processing by correcting background gradients aggravated by the low altitude of this object (see section "Final Cleanup" in Chap. 15) (Fig. 2.8).

Fig. 2.8. Thor's Helmet. North is up.

February: Clusters and Nebulae

Camera	ST10XME (H-alpha), ST2000XM (RGB)
Telescope	5.5-in. TEC refractor at f/7
Field of view	41 × 31 arcmin
Exposures	H-alpha 24 × 5 min, unbinned
	R 12 × 10 min, G and B each 6 × 10 min, unbinned
Scale	1.6 arcsec/pixel
Limiting magnitude	6.0

February 27: Medusa Nebula

Designations	Abell 21
Other names	PK 205+14.1
Right ascension	07 h 29.0 min
Declination	+13° 15'
Magnitude	13
Size	12 arcmin
Constellation	Gemini

The Medusa is an ancient planetary nebula, representing a dying star that has shed its outer layers, similar to the Dumbbell Nebula and Helix Nebula, but is more distant. It lies 1,000–1,500 light-years away in Gemini. The shell of hydrogen gas, glowing red, is energized by ultraviolet radiation from the central ember of the fading central star, appearing as a faint blue star in the center of the nebula on this image. The nebula's name is derived from its delicate filaments. The Medusa Nebula is so faint that it was not even discovered until 1955, by the Yerkes–McDonald survey (and named YM 29), and later designated as Abell 21.

Imaging. The Medusa Nebula is framed nicely by a medium focal length telescope with a field between 20 and 30 arcmin. It is moderately dim, but not as difficult to image as its listed magnitude of 13 might imply. From the suburbs, use an H-alpha or Red luminance to cut through light pollution and try to collect blue and green exposures at higher altitudes near the meridian. A single-shot color camera would have difficulty capturing this dim object from the suburbs. A dark sky is very helpful for imaging this object with any camera. Luminance layering can help gather the Medusa's light more efficiently.

Processing. Routine luminance layering techniques work well. For this image, the clear luminance exposures were short and therefore were combined with the red exposures for a stronger clear-red luminance. Inclusion of the clear luminance helped preserve some faint green components to the nebula's shell (Fig. 2.9).

February: Clusters and Nebulae

Fig. 2.9. Medusa Nebula. East is up.

Camera	ST2000XM
Telescope	5.5-in. TEC refractor at f/7
Field of view	39 × 31 arcmin
Exposures	Luminance clear 4 × 10 min, unbinned R 11 × 10 min, G 8 × 10 min, B 11 × 10 min, unbinned
Scale	1.6 arcsec/pixel
Limiting magnitude	6.0

February 27: Eskimo Nebula

Designation	NGC 2392
Other names	Caldwell 39, Clownface Nebula
Right ascension	07 h 29.2 min
Declination	+20° 55'
Magnitude	9.2
Size	0.7 arcmin
Constellation	Gemini

The Eskimo Nebula, NGC 2392, is another planetary nebula, residing 3,000 light-years away in the constellation Gemini. The colorful inner and outer layers have led some astronomers to call this the Clown Face Nebula, with the bright central star as its nose. Other astronomers see a man's head surrounded by a parka hood, and thus call this the Eskimo Nebula. In 2000, the Hubble Space Telescope imaged this nebula, discovering gas clouds so complex they are not yet understood. This is a young planetary nebula, created by gas released only a 1,000 years ago by a dying star. The inner filaments are being ejected by strong wind of particles from the central star.

Imaging. Despite its small size, the Eskimo Nebula can be a delight to image. Its small size concentrates its light into a small area, so that a large aperture is not essential. Use a long focal length telescope to yield a reasonably large image of the Eskimo, with an image scale less than 1 arcsec/pixel. Wait until a night of exceptional seeing, optimize your tracking, and focus carefully. If your tracking is less than perfect, shorten your exposures to as little as a minute or less, and take many more individual exposures.

Processing. Planetary photographers know that they can extract good images by taking many individual frames, and then selecting the best images for processing. For small planetary nebula, be brutal about rejecting individual exposures that show any blurring from atmospheric seeing or poor tracking. After combining your exposures into color channels, consider deconvolution to extract more detail from each channel. Then color combine, adjust your histogram with curves/levels, and sharpen the nebula even more aggressively. If noise from sharpening becomes excessive, soften the affect with a gentle Gaussian Blur. If the nearby eighth magnitude star is bloating, use star-shrinking techniques (Fig. 2.10).

Fig. 2.10. Eskimo Nebula. East is up.

February: Clusters and Nebulae

Camera	ST10XME
Telescope	12-in. Meade LX200R at $f/7$
Field of view	22 × 15 arcmin cropped to 11 × 8 arcmin
Exposures	R, G, and B each 6 × 5, unbinned
Scale	0.6 arcsec/pixel
Limiting magnitude	6.0

CHAPTER THREE

March: Clusters and Galaxies

March 1: Open Clusters M46 and M47

Designation	NGC 2437	NGC 2422	NGC 2423
Other names	Messier 46	Messier 47	
Right ascension	07 h 41.8 min	07 h 36.6 min	07 h 37.6 min
Declination	−14° 49′	−14° 29′	−13° 52′
Magnitude	6.1	4.4	4.4
Size	20 × 20 arcmin	25 × 25 arcmin	12 × 12 arcmin
Constellation	Puppis	Puppis	Puppis

M46 and M47 in the constellation Puppis are neglected by northern observers because they are so low in the horizon. Only for a few hours each night during the coldest winter months can these clusters be spotted, hovering above the trees. Like a pair of mismatched jewels, the faint 400 stars of M46 contrast with the bright but sparse 50 stars of M47. M46 (upper right) shows its age of 300 million years with some yellow suns at a distant 5,400 light-years, while younger M47 (lower left) is only 80 million years old and still has a few blue-white brilliant stars at 1,600 light-years. But there are three surprises in this view. In the lower right, the loose open cluster NGC 2423 can be defined from background stars. In the left center, a tiny cluster NGC 2425 can be seen. Best of all, superimposed on the north (right) edge of M46 is the 1.1-arcmin planetary nebula NGC 2438, which is just a foreground object only half the distance to M46.

Imaging. Open clusters can appear boring in images, especially if the field of view is so small that the cluster cannot be demarcated from the background. Generate more visual interest by framing this region with a short focal length telescope or a camera lens. But if your field is too large, the planetary nebula in M46 can become insignificant. Either a single-shot color camera or routine RGB imaging works best for open clusters. Luminance layering adds unneeded complexity for bright clusters. If your telescope has imperfect color correction, refocus between filter changes to keep stars sharp. If you use single-shot color, image close to the meridian to reduce blur from atmospheric refraction at low altitudes.

Processing. Routine image processing is sufficient. Concentrate on keeping star colors accurate. Large fields of view aggravate light-pollution

March: Clusters and Galaxies

Fig. 3.1. Open Clusters M46 (*top*), M47 (*lower left*), and NGC 2423 (*lower right*). East is up.

gradients. After adjusting your histogram and boosting color in the stars, subtract gradients until your background is smooth (see section "Dealing Out Gradients" in Chap. 15) (Fig. 3.1).

Camera	ST10XME
Telescope	3.5-in. Takahashi refractor at f/4.5
Field of view	126 × 85 arcmin
Exposures	R, G, and B each 10 × 3 min, unbinned
Scale	3.5 arcsec/pixel
Limiting magnitude	6.0

March: Clusters and Galaxies

March 2: Spiral Galaxy NGC 2403

Designations	NGC 2403
Other names	Caldwell 7
Right ascension	07h 36.9m
Declination	+65° 36'
Magnitude	8.2
Size	23 × 12 arcmin
Constellation	Camelopardalis

Spiral Galaxy NGC 2403 is one of the largest and brightest galaxies that were overlooked by Charles Messier in his famous catalog. Although neglected by Messier, NGC 2403 has been recognized as entry number 7 in Patrick Moore's Caldwell catalog. At a distance of 12 million light-years in the faint constellation Camelopardalis (the giraffe), it lies in the same galaxy group as the better known M81 and M82 galaxies. The red areas are bright emission nebulae of glowing hydrogen, and the hazy blue areas are star clusters of young blue-white stars. Two bright stars overlie the 3:00 and 8:00 positions of the galaxy in this image, and are not part of NGC 2403, but instead are foreground stars within our own galaxy.

Imaging. NGC 2403 is a showpiece for deep sky imaging. The galaxy is almost as large and bright as M81, and can be framed nicely with either middle or long focal lengths. NGC 2403 has abundant H-II regions and a detailed core. To capture this detail, accurate tracking, steady skies, and sharp focus are important. The galaxy is bright enough for single-shot color cameras or routine RGB methods. Luminance layering helps to collect light more efficiently, which allows extraction of more detail during processing. Obtain your luminance frames when the skies are most steady and obtain your binned RGB frames when skies are more turbulent.

Processing. NGC 2403 is bright, so your imaging data should be rich. Routine LRGB (or stacking for single shot cameras) methods work well. The core and inner arms are bright and detailed, and can tolerate some aggressive sharpening if your data has a good signal-to-noise ratio. Enhance your color with either saturation, match color, or LAB model techniques to reveal the plentiful H-II regions and rich star clusters in the spiral arms (see section "Color Enrichment" in Chap. 15) (Fig. 3.2).

Fig. 3.2. Galaxy NGC 2403 (Caldwell 7). North is up.

March: Clusters and Galaxies

Camera	ST10XME
Telescope	12-in. Meade LX200R at $f/7$
Field of view	23 × 15 arcmin
Exposures	Luminance clear 48 × 5 min, unbinned R and G each 18 × 5 min, B 22 × 5 min, binned 2 × 2
Scale	0.6 arcsec/pixel
Limiting magnitude	6.0

March 20: Ancient Open Cluster M67

Designation	NGC 2682
Other names	Messier 67
Right ascension	08 h 51.3 min
Declination	+11° 48'
Magnitude	6.9
Size	30 × 30 arcmin
Constellation	Cancer

Open Cluster M67 contains about 500 stars at a distance of 2,700 light-years, in the direction of constellation Cancer. M67's age of over 3 billion years is much older than most other open clusters. Usually, the weak gravitational attraction of the loosely grouped open clusters allows them to become disrupted by interaction with other stars, clusters, and interstellar gas and dust. Thus, most open clusters only survive intact for a few hundred million years and so remain dominated by young bright blue-white stars, which burn out faster than dimmer yellow and red suns. In contrast, this image of M67 shows mostly white and yellow suns, indicative of its greater age. It is unclear why M67 has survived so long, but some have suggested that because M67 is slightly above the galactic plane, it may have avoided the types of gravitational interactions that disperse most open clusters.

Imaging. Frame M67 to allow its concentration of stars to be distinguished from the background distribution of stars. My field of view is about the minimum size to define M67 as a cluster. Keep exposures short enough to avoid excessive blooming from the two stars of magnitude 7.8 in the northern portions of the cluster. As with most bright open clusters, luminance layering is counterproductive. Routine RGB or single-shot color techniques work well. If you are using a single-shot color camera, try to image when the object is near the meridian, to avoid atmospheric refraction at lower altitudes that can smear your colors. If you are using RGB filters, remember to refocus between filter changes, especially if your telescope has color correction that is less than perfect.

Processing. Alignment of images is critical for star clusters to avoid asymmetric color halos. After enriching color, confirm correct color balance.

March: Clusters and Galaxies

Fig. 3.3. Open Cluster M67. Northeast is up.

This cluster should appear more yellow and less blue than most other clusters (Fig. 3.3).

Camera	ST10XME
Telescope	5.5-in. TEC refractor at f/7
Field of view	52 × 35 arcmin
Exposures	R, G, and B each 8 × 3 min, unbinned
Scale	1.4 arcsec/pixel
Limiting magnitude	3.5

March 30: Barred Spiral Galaxy NGC 2903

Designations	NGC 2903
Other names	Hidden Treasure 51
Right ascension	09 h 32.2 min
Declination	+21° 30'
Magnitude	8.9
Size	13 × 7 arcmin
Constellation	Leo

Galaxy NGC 2903 lies 20 million light-years away within the Leo Spur of Galaxies. The galaxy is tilted 24° from edge-on, allowing us to see its spiral structure. The inner core has a prominent bar, oriented vertically in this image. The barred spiral galaxies often are associated with high rates of star formation, termed "hot spot galaxies," but not quite at the rate of the starburst galaxies like M77 and M82. The gravitational pull of the central bar draws material to the galaxy's core, accelerating star formation. Large H-II regions, appearing red in this image, join a ring of active star formation around the core. Farther out, two dominant spiral arms extend outward, shining blue with young star clusters.

Imaging. The galaxy is oriented close to the north–south axis, but would appear more dynamic if framed at an angle of 20–30° from the long axis of the image. Use a long focal length to extract detail in this image, but this also requires good tracking, sharp focus, and steady skies. LRGB data acquisition allows more efficient collection of light data. If possible, collect your luminance data when the skies are steadiest and when the galaxy is at its highest altitude near the meridian.

Processing. Routine luminance layering methods work best. Apply digital development or curves/levels carefully to detect the outer symmetrical spiral arms without burning out the core. Sharpen the central galaxy to reveal the bright bar extending south from the core, and the texture in the spiral arms. Enrich color to emphasize H-II regions in the spiral arms (Fig. 3.4).

Fig. 3.4. Galaxy NGC 2903. North is up.

March: Clusters and Galaxies

Camera	ST10XME
Telescope	12-in. Meade LX200R at $f/7$
Field of view	23×15 arcmin
Exposures	Luminance clear 18×5 min, unbinned R and B each 6×5 min, G 4×5 min, binned 2×2
Scale	0.6 arcsec/pixel
Limiting magnitude	6.0

CHAPTER FOUR

April: Galaxy Pairs and Groups

April 5: Galaxies M81 and M82

Designation	NGC 3031	NGC 3034
Other names	Messier 81, Bode's Galaxy	Messier 82, Cigar Galaxy
Right ascension	09 h 55.8 min	09 h 56.2 min
Declination	+69° 04'	+69° 42'
Magnitude	6.9	8.4
Size	26 x 14 arcmin	11 x 5 arcmin
Constellation	Ursa Major	Ursa Major

April is the month for interacting galaxies, beginning with M81 (lower left) and M82 (upper right). These two galaxies, about 12 million light-years distant, passed close to one another about 20 million years before the light began its journey to my camera. Obviously, the galaxies are now farther apart than this image shows, but we would not see that for another 12 million years…confused yet? During this close encounter event, the larger and more massive M81 dramatically deformed M82 by gravitational interaction, stimulating a burst of star formation. The galaxies are still close together, with their centers separated by a linear distance of only about 150,000 light-years.

Imaging. M81 and M82 can be imaged either together, as shown on the wide-field view, or individually, as shown on the next two pages. H-II regions have been emphasized in M82 by enriching the red channel with H-alpha data. Despite its large size, M81 shows little detail in its spiral arms unless long exposures are obtained away from light pollution. M82 is better framed with a smaller field of view. Both objects are bright enough for routine RGB methods or single-shot color cameras, but obtaining some data with H-alpha filters aids detection of the brilliant H-II glow around the core of M82.

Processing. Routine luminance layering methods can be enhanced with H-alpha data to highlight the starburst region in M82. After routine image processing, enrich color either by boosting saturation or by using Photoshop's match color. For images of M81 alone, you can boost contrast between the yellow core and blue spiral arms in Photoshop by converting mode to Lab color, selecting the "b" channel, adjusting contrast upward, then converting back to RGB (Fig. 4.1).

April: Galaxy Pairs and Groups

Fig. 4.1. (a) Galaxies M81 and M82. North-northeast is up.

Fig. 4.1. (**b**) Galaxy M81. North is up.

April: Galaxy Pairs and Groups

Fig. 4.1. (c) Galaxy M82. North-northeast is up.

Camera (M81 and M82)	ST10XME
Telescope	5.5-in. TEC refractor at *f*/5
Field of view	73 × 49 arcmin
Exposures	Luminance clear 12 × 5 min, unbinned R, G, and B each 6 × 5 min, Hα 8 × 5 min, binned 2 × 2
Scale	2.0 arcsec/pixel
Limiting magnitude	6.0
Camera (M81)	ST2000XM
Telescope	5.5-in. TEC refractor at *f*/7
Field of view	42 × 31 arcmin cropped to 38 × 28 arcmin
Exposures	Luminance clear 8 × 10 min, unbinned R 10 × 10 min, G and B each 5 × 10 min, unbinned
Scale	1.6 arcsec/pixel
Limiting magnitude	6.0
Camera (M82)	ST10XME
Telescope	12-in. Meade LX200R at *f*/7
Field of view	22 × 15 arcmin
Exposures	H-alpha 8 × 5 min, binned 2 × 2 R and G each 8 × 5 min, B 7 × 5 min, unbinned
Scale	0.6 arcsec/pixel
Limiting magnitude	6.0

April 10: Little Pinwheel Galaxy

Designations	NGC 3184
Other names	Hidden Treasure 52
Right ascension	10 h 18.3 min
Declination	+41° 25'
Magnitude	9.6
Size	7 × 7 arcmin
Constellation	Ursa Major

Galaxy NGC 3184 is seen only 20° from face on, providing a view resembling the Pinwheel Galaxy M101 in the same constellation. Both galaxies are about 60,000 light-years in diameter, but NGC 3184 is nearly twice as distant. In 1999, a supernova was observed in NGC 3184 and allowed a calculation of the galaxy's distance at 36 million light-years. The yellow core harbors old mature stars, and the spiral arms show several red H-II regions amid clusters of young blue-white stars. In the upper right of this image, a red giant star of the spectral class M2III with a magnitude of 6.5 can be found about 10 arcmin to the west of NGC 3184.

Imaging. Frame the Little Pinwheel Galaxy with a small field of view about 20 arcmin across. You may wish to include the nearby red giant star at the other side of your field of view to provide visual balance and contrast with the galaxy. Including this star is easier with an antiblooming camera that will restrain the star from bloating excessively in your long exposures for the galaxy. As with most dim galaxies, LRGB methods work well, and allow a nice image to be obtained in about 3 h of exposures. Single-shot color cameras or routine RGB would require longer exposure times to capture this much detail.

Processing. After calibrating your exposures, apply deblooming tools if needed for the red giant star to the west of NGC 3184. Luminance layering methods are suggested. If you have long, deep clear luminance exposures, try deconvolution after creating your luminance channel. After routine histogram adjustments and color enhancement, sharpen the central regions of the galaxy and smooth the dimmer regions of the image. If your image includes the nearby bright red star, apply cleanup techniques to tone down bloating of this star (Fig. 4.2).

Fig. 4.2. Little Pinwheel Galaxy NGC 3184. Northwest is up.

April: Galaxy Pairs and Groups

Camera	ST10XME
Telescope	12-in. Meade LX200R at f/7
Field of view	22 × 15 arcmin cropped to 20 × 13 arcmin
Exposures	Luminance clear 18 × 5 min, unbinned R 6 × 5 min, G 4 × 5 min, B 10 × 5 min, binned 2 × 2
Scale	0.6 arcsec/pixel
Limiting magnitude	6.0

April 11: Hickson 44 Galaxy Group

Designations	NGC 3190	NGC 3193	NGC 3187	NGC 3185
Other names	Hickson 44, Arp 316	Hickson 44, Arp 316	Hickson 44, Arp 316	Hickson 44, Arp 316
Right ascension	10 h 18.1 min	10 h 18.4 min	10 h 17.8 min	10 h 17.6 min
Declination	+21° 50'	+21° 53'	+21° 52'	+21° 41'
Magnitude	11.1	11.2	12.7	12.0
Size	5×2 arcmin	3×3 arcmin	3×1 arcmin	2×2 arcmin
Constellation	Leo	Leo	Leo	Leo

At a distance of 60 million light-years in the constellation Leo, the Hickson 44 Galaxy Group contains four galaxies in a tight group that showcase the diversity of galaxies. At the upper left, the elliptical galaxy NGC 3193 appears as an amorphous conglomeration of stars, lacking any spiral structure. Elliptical galaxies are relatively inert, composed largely of older stars, and lack sufficient interstellar gas or dust to undergo much new star formation. Therefore, elliptical galaxies also lack the red glowing hydrogen-alpha regions, the blue-white clusters of newly formed stars, and dark dust lanes that enrich images of spiral galaxies. The upper center of the image is the tightly wound spiral galaxy NGC 3190, tilted only 8° from edge on. Its central dust lane is warped by gravitational interaction with nearby galaxies. Right of center is the spiral galaxy NGC 3187, with gravitational forces stretching the edges of its spiral arms into tidal tails. In the lower right is the barred spiral galaxy NGC 3185.

Imaging. The individual galaxies are small and therefore best framed as a group. Two framing options are suggested. The larger frame is shown, including all four galaxies. If your imaging chip is smaller, a tighter frame of NGC 3190, 3193, and 3187 can be successful with a field of 10×15 arcmin.

Processing. Routine luminance layering methods work best. Try to remove blooms around the brighter stars before aligning and combining. After you finish your histogram adjustments and color enhancement, deal with any gradients that may degrade the large spaces between the galaxies. Finally, shrink bloated stars, especially the star with a magnitude of 7.6 in the upper right, which might otherwise be confused with elliptical galaxies (see section "Final Cleanup" in Chap. 15) (Fig. 4.3).

April: Galaxy Pairs and Groups

Fig. 4.3. Hickson 44 Galaxy Group. North is up.

Camera	ST10XME
Telescope	12-in. Meade LX200R at *f*/7
Field of view	22 × 15 arcmin
Exposures	Luminance clear 25 × 5 min, unbinned R 8 × 5 min, G 7 × 5 min, B 10 × 5 min, binned 2 × 2
Scale	0.6 arcsec/pixel
Limiting magnitude	6.0

April 17: Galaxy Pair M95 and M96

Designation	NGC 3351	NGC 3368
Other names	Messier 95	Messier 96
Right ascension	10 h 44.0 min	10 h 46.8 min
Declination	+11° 42'	+11° 49'
Magnitude	9.7	9.2
Size	7 × 5 arcmin	7 × 5 arcmin
Constellation	Leo	Leo

Galaxies M95 and M96 are a pair of spiral galaxies within the Leo I group of galaxies 38 million light-years away. Both were discovered by Pierre Mechain on March 20, 1781, and were added to Charles Messier's famous catalog 4 days later. M95 (lower right) is a "barred" spiral galaxy, with a bar shape at its core and discrete spiral arms. M96 (upper left) is slightly brighter, containing a bright inner disk of old yellow stars, an inner ring of younger blue-white stars, surrounded by a fainter outer ring also dominated by younger stars.

Imaging. You can choose a much wider frame that includes the elliptical galaxies M105 and NGC 3384, but this would require a field of at least 90 × 60 arcmin, which would make the galaxies appear quite small and sparse. The frame in my image is the minimum field to capture both M95 and M96 without the galaxies crowding the edges of the image. Individually, M95 and M96 are both small and difficult to image with amateur telescopes, and appear more interesting as a pair.

Processing. Routine LRGB works well for these galaxies. Because of the large open space between the galaxies, the stars become an important element of the composition. Precise alignment of the color channels is critical to keep star colors centered without asymmetric halos. Sharpen the brighter regions in the galaxies. Try to minimize sharpening of stars that may brighten the stars and distract from the galaxies. Try to avoid excess color in the stars (Fig. 4.4).

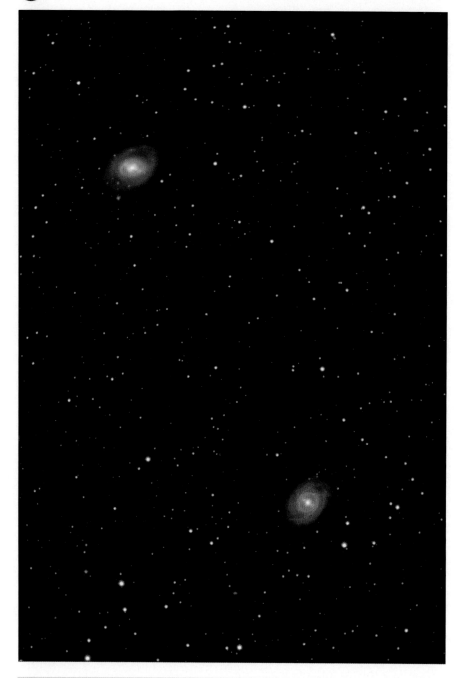

Fig. 4.4. Galaxies M96 (*upper left*) and M95 (*lower right*). Northeast is up.

April: Galaxy Pairs and Groups

Camera	ST10XME
Telescope	5.5-in. TEC refractor at f/5
Field of view	72 × 48 arcmin
Exposures	Luminance clear 14 × 5 min, unbinned R, G, and B each 9 × 5 min, Ha 8 × 5 min, binned 2 × 2
Scale	2.0 arcsec/pixel
Limiting magnitude	6.0

April 24: Owl Nebula M97 with Galaxy M108

Designation	NGC 3587	NGC 3556
Other names	Messier 97	Messier 108
Right ascension	11 h 14.9 min	11 h 11.6 min
Declination	+55° 01'	+55° 40'
Magnitude	9.9	10.0
Size	3 × 3 arcmin	8 × 3 arcmin
Constellation	Ursa Major	Ursa Major

In the lower left of this image, the Owl Nebula is the ghost-like outer shell of gas blown off by a dying star, at a distance of about 3,000 light-years. Discovered in 1781, even small telescopes can detect the faint hollow eyes of the nebula, which earned its name in 1848. In the upper right of this image, Galaxy M108 lies over a 1,000 times farther away, at a distance of 45 million light-years. Its spiral form is concealed by its nearly edge-on orientation.

Imaging. The Owl Nebula appears more interesting when paired with M108. The highest resolution amateur images of the Owl Nebula lack much more detail than in this image, because the shells of gas are diffuse and ill-defined. A slightly larger field of 75 × 50 arcmin would frame these objects nicer, with less crowding at the edges of the frame. With a large field of view, both the Owl and M108 will concentrate their tenth magnitude light over fewer pixels, so that long exposures are not essential. Single-shot color, RGB color, or LRGB are all options for these two objects with a large field of view.

Processing. Routine processing techniques can be used for these objects. After calibrating your images, remove blooms from the bright stars, especially the star with a magnitude of 7.7 midway between the Owl and M108. Align your stars carefully to avoid distracting asymmetric color halos. After histogram adjustments, sharpen the Owl to show its eyes, and M108 to show its dark lanes. Enhance colors to reveal a rich teal center and red rim in the Owl. Correct any gradients that may distract the viewer in the broad region between the two objects. Finally, shrink halos around the brighter stars (Fig. 4.5).

April: Galaxy Pairs and Groups

Fig. 4.5. Owl Nebula M97 (*lower left*) and Galaxy M108 (*upper right*). North is up.

Camera	ST10XME
Telescope	5.5-in. TEC refractor at f/5
Field of view	71 × 48 arcmin
Exposures	Luminance IDAS 31 × 5 min, unbinned R and G each 4 × 5 min, B 6 × 5 min, binned 2 × 2
Scale	2.0 arcsec/pixel
Limiting magnitude	3.5

April 26: Galaxy Trio in Leo

Designation	NGC 3623	NGC 3627	NGC 3628
Other names	Messier 65	Messier 66	Arp 317 group
Right ascension	11 h 18.9 min	11 h 20.3 min	11 h 20.3 min
Declination	+13° 06′	+13° 00′	+13° 35′
Magnitude	9.3	8.9	9.5
Size	10 × 3 arcmin	9 × 4 arcmin	14 × 3 arcmin
Constellation	Leo	Leo	Leo

In the constellation Leo the Lion, a group of distant galaxies stands out supreme known as the Galaxy Trio in Leo. In the 1960s, Caltech astronomer Halton Arp included this galaxy group in his famous Atlas of Peculiar Galaxies as entry Arp 317. These galaxies lie at a distance of 35 million light-years. M65 is at the lower right of this image with tight spiral arms, similar to our own Milky Way Galaxy. M66, at the lower left, has had its spiral arms distorted, probably due to interaction with M65. NGC 3828, at the top, is an "edge-on" spiral galaxy, with a broad band of interstellar dust obscuring the central disk. For an inhabitant of any of these galaxies, the views of the other galaxies would be spectacular, covering as much of the sky as one of our constellations!

Imaging. The Leo Trio is one of the most popular targets for deep sky imaging. A frame of 60 × 40 arcmin captures all three galaxies beautifully. A smaller field of 40 × 30 arcmin can be used to frame just M65 and M66. High-resolution imaging of any of these objects individually with a field of 20 × 15 arcmin is a challenge for the more experienced imager (see April 26: Hamburger Galaxy NGC 3628). When imaging the entire Leo Trio, minimize gradients in light-polluted suburbs by obtaining images near the meridian when these objects are at their highest altitude. Large-field imaging of the Leo Trio can be successful with single-shot color, RGB imaging, or LRGB imaging. The faint edges of NGC 3628 will benefit most from LRGB technique.

Processing. The suggestions outlined for imaging the Owl Nebula and M108 together (section "April 24: Owl Nebula M97 with Galaxy M108) also work well for this galaxy group. These methods seek to avoid distractions from stars and gradients (Fig. 4.6).

Fig. 4.6. Galaxy Trio in Leo. North is up.

April: Galaxy Pairs and Groups

Camera	ST10XME
Telescope	5.5-in. TEC refractor at $f/7$
Field of view	52 × 35 arcmin
Exposures	Luminance IDAS 36 × 5 min, unbinned R and G each 8 × 5 min, B 12 × 5 min, binned 2 × 2
Scale	1.4 arcsec/pixel
Limiting magnitude	3.5

April 26: Hamburger Galaxy NGC 3628

Designation	NGC 3628
Other names	Arp 317, Hidden Treasure 58
Right ascension	11 h 20.3 min
Declination	+13° 35'
Magnitude	9.5
Size	14 × 3 arcmin
Constellation	Leo

Galaxy NGC 3628 is the largest but dimmest member of the famous Galaxy Trio in Leo, joining galaxies M65 and M66 (see section "April 26: Galaxy Trio in Leo"). NGC 3628 is seen edge on, with a broad dust lane crossing the central bulge. Gravitational interactions between these galaxies warp of the ends of NGC 3628, and deform the central bulge. As a consequence, NGC3628 lacks the smooth symmetry of other edge-on spirals like the Needle Galaxy (NGC 4565), the Sombrero Galaxy (M104), and the Outer Limits Galaxy (NGC 891).

Imaging. NGC 3628 harbors both incredible detail in its core and faint featureless warped outer arms. Frame this object generously around the arms to avoid amputating the edges of the galaxy. LRGB methods work best for this object, with as much luminance data as you are willing to collect. For this image, the larger telescope was devoted solely to the collection of a clear luminance. RGB data was collected with a less-efficient camera on a smaller telescope to avoid diverting any imaging time away from the luminance channel.

Processing. Hybrid LRGB is an option for combining luminance data from one telescope with RGB data from another telescope. The RGB data is processed routinely to the level of the RGB combine. The combined RGB image is then aligned and scaled to the combined luminance image. This scaling can be accomplished in Maxim DL, CCDStack, RegiStar, or some other astronomical imaging programs. After this alignment, routine luminance layering can be employed. Adjust the histogram to reveal the faint outer arms. Sharpen the central dark lane. Correct color balance and then boost the color. Finally, tone down halos around the two stars with a magnitude of 10 to the north and south of NGC 3628 (Fig. 4.7).

April: Galaxy Pairs and Groups

Fig. 4.7. Hamburger Galaxy NGC 3628. Northwest is up.

Cameras	ST10XME (luminance), ST2000XM (RGB)
Telescopes	12-in. Meade LX200R at f/7 (luminance) 5.5-in. TEC refractor at f/7 (RGB)
Field of view	22 × 15 arcmin
Exposures	Luminance clear 37 × 5 min, unbinned R 6 × 10 min, G 4 × 10 min, B 10 × 10 min, unbinned
Scale	0.6 arcsec/pixel
Limiting magnitude	6.0

April 30: Galaxy Pair NGC 3718 and NGC 3729

Designation	NGC 3718	NGC 3729
Other names	Arp 214	
Right ascension	11 h 32.6 min	11 h 33.8 min
Declination	+53° 04'	+53° 08'
Magnitude	10.5	11.6
Size	8 × 4 arcmin	3 × 2 arcmin
Constellation	Ursa Major	Ursa Major

Galaxy NGC 3718, at the lower left of this image, is entry #214 in the Arp catalog of Peculiar Galaxies published in 1966. The most striking feature of NGC 3718 is the "warped" dust lane running through the stellar bulge. This dust lane is visible because the galaxy is almost edge-on to our line of sight. The shape seen here, however, differs dramatically from other edge-on galaxies like NGC 891 or NGC 4565. Because we are seeing the galaxy from the side, the curve of the dust filaments at the top and bottom of the galaxy are not simply "spiral" arms, but rather "warping" of the galaxy's arms above and below the plane of the galaxy. Why? Gravitational interaction with galaxy NGC 3729, only 150,000 light years away, is contributing to this distortion.

Imaging. Although NGC 3718 is a small object, its peculiar shape and interaction with NGC 3729 justify its inclusion in the list of 100 Best Targets. A slightly larger field of 25 × 20 arcmin would have allowed inclusion of a small galaxy group at the southern edge of NGC 3718. High resolution is helpful to extract fine detail from NGC 3718. This requires the combination of accurate tracking, sharp focusing, and steady atmospheric seeing. Because these objects are faint, LRGB methods are preferred to gather light more rapidly.

Processing. Routine luminance layering methods are recommended. To extract as much detail as possible, perform deconvolution on your combined luminance image prior to application of either DDP or curves/levels. Later, devote extra time to sharpening the core of NGC 3718. This can be accomplished in two or more steps, first with a radius of only several pixels to sharpen the thin dark lane, and then with larger radius to sharpen the larger features. Try to avoid erasing the faint outer arms when dealing with any gradients in your image (Fig. 4.8).

Fig. 4.8. Galaxy Pair NGC 3718 (*lower left*) and NGC 3729 (*upper right*). East is up.

April: Galaxy Pairs and Groups

Camera	ST10XME
Telescope	12-in. Meade LX200R at $f/7$
Field of view	22 × 15 arcmin
Exposures	Luminance clear 31 × 5 min, unbinned R 12 × 5 min, G 9 × 5 min, B 12 × 5 min, binned 2 × 2
Scale	0.6 arcsec/pixel
Limiting magnitude	6.0

CHAPTER FIVE

May: Diversity of Galaxy Shapes

May 6: Galaxy M109

Designation	NGC 3992
Other names	Messier 109
Right ascension	11 h 57.7 min
Declination	+53° 22'
Magnitude	9.8
Size	8 × 5 arcmin
Constellation	Ursa Major

Galaxy M109 is a stunning example of a "barred spiral galaxy." This category of spiral galaxies is defined by a distinct bar of densely packed stars elongating the nucleus of the galaxy. The spiral arms originate from the ends of this bar. M109 is the brightest of 50 galaxies in the M109 Group, a large group of galaxies located in the constellation Ursa Major at a distance of 55 million light-years.

Imaging. Although relatively small among the galaxies chosen for this book, M109 concentrates its photons of magnitude 9.8 into a bright core and crisp spiral arms. The best detail is extracted with long exposures and luminance layering, but nice results can be obtained from the suburbs with as little as 2 h of exposures. Use a long focal length to obtain high resolution and a frame of around 20 arcmin or less. As with other small objects, you will gather the sharpest details under steady skies, with accurate tracking and careful focus.

Processing. If you have rich data in your luminance, apply deconvolution after aligning and combining your luminance images. Then, after either digital development or curves/levels, sharpen the core and brighten the spiral arms with your favorite tools. When sharpening in Photoshop, deselect the center of the core and feather the selection a few pixels, or you may get an unnatural round bright circle at the core. Also, deselect the 13th magnitude foreground star in the north side of M109 when sharpening. Consider sharpening in two phases, each gently. The first, with a radius of 5–8 pixels, is to enrich the small scale contrast within the spiral arms. The second, with a radius of 15–20 pixels, is to enhance the borders of the bar and spiral arms by dimming the dark lanes. Finally, shrink any excessive bloating in the ninth magnitude star just to the southwest of M109 (Fig. 5.1).

May: Diversity of Galaxy Shapes

Fig. 5.1. Galaxy M109. East-Northeast is up.

Camera	ST10XME
Telescope	12-in. Meade LX200R at f/7
Field of view	23 × 15 arcmin
Exposures	Luminance clear 45 × 5 min, unbinned
	R, G, and B each 5 × 5 min, binned 2 × 2
Scale	0.6 arcsec/pixel
Limiting magnitude	6.0

May 11: Silver Needle Galaxy

Designations	NGC 4244
Other names	Caldwell 26
Right ascension	12 h 17.5 min
Declination	+37° 49'
Magnitude	10.2
Size	16×2 arcmin
Constellation	Canes Venatici

The Silver Needle Galaxy is a small spiral galaxy seen exactly edge-on. Its size of 46,000 light-years is almost the same as the Pinwheel Galaxy M33, but the Silver Needle is about four times farther away at a distance of 10 million light-years. The Gemini Observatory has recently reported that a massive nuclear stellar cluster, similar in size to a globular cluster, is rotating strongly at the core of the galaxy.

Imaging. The Silver Needle Galaxy has a length of 16 arcmin, and thus can be framed nicely with a field of view between 20 and 40 arcmin. Although the Silver Needle's magnitude of 10.2 may seem out of reach for small telescopes, the edge-on orientation concentrates its light onto a thin long sliver. My image was obtained despite a thin haze and a nearly full moon. Even from the suburbs with a 5.5-in. telescope, I have been able to capture nice detail in just 3 h with luminance layering. RGB imaging and single-shot color are possible but would require longer exposures to obtain comparable detail.

Processing. The Silver Needle Galaxy has less detail in its dark lane than either NGC 891 or NGC 4565, so do not be disappointed when your sharpening methods seem to show limited effect. Avoid introducing too much noise that can detract from the simple elegance of this thin sliver of a galaxy. This galaxy also has subdued colors, perhaps contributing to its name as the Silver Needle. Avoid overemphasizing the colors during processing. If your field of view is over 30 arcmin, consider shrinking the several bright stars, up to magnitude 9.4 to the south of the galaxy, that may bloat and distract from your primary target (Fig. 5.2).

Fig. 5.2. Silver Needle Galaxy. North is up.

May: Diversity of Galaxy Shapes

Camera	ST10XME
Telescope	12-in. Meade LX200R at $f/7$
Field of view	22 × 15 arcmin
Exposures	Luminance clear 14 × 5 min, unbinned
	R, G, and B each 6 × 5 min, binned 2 × 2
Scale	0.6 arcsec/pixel
Limiting magnitude	4.0

May 11: Galaxy M106

Designation	NGC 4528
Other names	Messier 106
Right ascension	12 h 19.0 min
Declination	+47° 18'
Magnitude	8.4
Size	18 × 8 arcmin
Constellation	Canes Venatici

Galaxy M106 lies at a distance of 25 million light-years. This galaxy was discovered by Pierre Mechain, a protégé of Charles Messier, in 1781. Mechain's discoveries during this decade subsequently were appended to Messier's catalog as numbers M102–M110. The yellow regions in the central spiral arms arise from mature stars billions of years old, whereas the faint blue outer spiral arms announce younger and brighter stars only several millions of years old.

Imaging. M106 is a delightful galaxy for imaging because it is large, bright, colorful, and rich in detail. A larger frame might improve the space around the faint outer arms, and would fully include the small 3 × 1 arcmin spiral galaxy at the northwest corner of this image. The frame could also be rotated slightly to keep M106 from following the long axis of the frame too precisely, which creates a static posed appearance in my image. LRGB methods help to reveal the faint outer arms, but acceptable results can be achieved with single-shot color and long exposures. The longer your exposures, the better you will able to bring in some detail to the faint outer arms when you process your image.

Processing. After adjusting your histogram with DDP or levels/curves, sharpen the bright central core aggressively to reveal the dark brown channels of dust. After balancing color in your RGB channels, enhance the color to emphasize the yellow core and the blue outer arms. This can be achieved by either increasing saturation or match color adjustments. Another option that works well for M106 is to convert your image to a Lab color model, as discussed in the section "Color Enrichment" in Chap. 15. Finally, apply some blurring (or Photoshop's reduce noise filter) to the faint outer arms to smooth away noise and grainy color (Fig. 5.3).

May: Diversity of Galaxy Shapes

Fig. 5.3. Galaxy M106. North–northwest is up.

Camera	ST10XME
Telescope	12-in. Meade LX200R at *f*/7
Field of view	23 × 15 arcmin
Exposures	Luminance clear 37 × 5 min, unbinned R and G each 8 × 5 min, B 9 × 5 min, binned 2 × 2
Scale	0.6 arcsec/pixel
Limiting magnitude	6.0

May 12: Galaxies M100 and NGC 4312

Designation	NGC 4321	NGC 4312
Other names	Messier 100	
Right ascension	12 h 22.9 min	12 h 22.5 min
Declination	+15° 49′	+15° 32′
Magnitude	9.3	12.1
Size	7 × 6 arcmin	5 × 1 arcmin
Constellation	Coma Berenices	Coma Berenices

Galaxy M100, a member of the Virgo Cluster of Galaxies, resides at a distance of 50 million light-years in the constellation Coma Berenices. The galaxy extends out quite far beyond the spiral arms…the faint glow is real and is not a processing artifact. M100 is surrounded by several small dwarf galaxies. At the lower right is NGC 4312, a smaller companion spiral galaxy seen end-on.

Imaging. You have several options for framing M100. The most pleasing images include both M100 and NGC 4312, which benefit from a frame of at least 35 × 25 arcmin. My frame could be tighter, but this would also remove some of the faint background galaxies that add extra interest. Avoid lining up the two galaxies in the long axis of your photo; an angle between the objects keeps the view dynamic. Keeping the field of view wider also reduces the demands for perfect tracking and steady skies. Luminance layering allows more rapid imaging of this pair, but they are sufficiently bright for routine RGB or single-shot color cameras.

Processing. Routine image processing is suggested. Sharpen M100 to emphasize the spiral arms. Minimally sharpen NGC 4312 to avoid unnatural artifacts. If you are removing gradients, be careful that you do not erase the half dozen background galaxies that populate this field. If the star with a magnitude of 9.7 midway between M100 and NGC 4312 is bloated or otherwise detracting, apply star-shrinking methods (see section "Final Cleanup" in Chap. 15) (Fig. 5.4).

Fig. 5.4. Galaxies M100 and NGC 4312. North is up.

May: Diversity of Galaxy Shapes

Camera	ST10XME
Telescope	5.5-in. TEC refractor at $f/7$
Field of view	51 × 35 arcmin cropped to 44 × 30 arcmin
Exposures	Luminance IDAS 21 × 5 min, unbinned R 9 × 5 min, G 7 × 5 min, B 9 × 5 min, binned 2 × 2
Scale	1.4 arcsec/pixel
Limiting magnitude	6.0

May 13: Markarian's Chain

Designations	NGC 4406	NGC 4374	NGC 4438,4435
Other names	Messier 86	Messier 84	The Eyes
Right ascension	12 h 26.2 min	12 h 25.1 min	12 h 27.8 min
Declination	+12° 57'	+12° 53'	+13° 01'
Magnitude	9.0	9.4	10.0, 10.8
Size	9 × 6 arcmin	7 × 6 arcmin	8 × 3, 3 × 2 arcmin
Constellation	Virgo	Virgo	Virgo

The Virgo Galaxy Cluster contains thousands of individual galaxies centered at a distance of about 60 million light-years, occupying over 10° of the sky. The Virgo Galaxy Cluster lies at the center of the larger Virgo Supercluster, which includes our Local Group. Our Local Group is a smaller galaxy cluster that includes the Milky Way, Andromeda, Pinwheel, M81, and M82 galaxies.

Visual astronomers are often drawn to an arc of galaxies in the heart of the Virgo cluster, called Markarian's Chain, which stretches from M84 to NGC 4477. This image of the western half of Markarian's Chain is centered on the giant elliptical galaxy M86. Another bright elliptical galaxy M84 is on the lower right of this image. In the upper left of the image, two interacting galaxies are seen called "The Eyes" because of their appearance through the eyepiece of a telescope. Edge-on spiral galaxies around M86 include NGC 4402 to its right and NGC 4388 to its lower left. Many other smaller galaxies can be seen in this image.

Imaging. The Virgo Galaxy Cluster can be framed in many ways. One of the most popular is shown here, anchored by M86 near center, and flanked by M84 on one side and by The Eyes on the other. A much wider view of 2° can be extended to include M87 to the southeast and NGC 4459 to the northeast. On a night of perfect seeing, you can try to image The Eyes in high resolution.

Processing. Because so much of the field is empty space, make sure your stars are aligned well to avoid oval stars and distracting asymmetric color halos. Focally sharpen the spiral galaxies. Correct any distracting background gradients (Fig. 5.5).

May: Diversity of Galaxy Shapes

Fig. 5.5. Western portion of Markarian's Chain. Northeast is up.

Cameras	ST10XME (luminance), ST2000XM (color)
Telescopes	5.5-in. TEC refractor at f/5 (luminance)
	4-in. Astro-Physics refractor at f/6 (color)
Field of view	66 × 48 arcmin
Exposures	Luminance IDAS 16 × 10 min, unbinned
	R 6 × 10 min, G and B 3 × 10 min, unbinned
Scale	2.0 arcsec/pixel
Limiting magnitude	3.5

May 16: Needle Galaxy

Designation	NGC 4565
Other names	Caldwell 38
Right ascension	12 h 36.3 min
Declination	+25° 59'
Magnitude	9.6
Size	16 × 3 arcmin
Constellation	Coma Berenices

The spectacular spiral galaxy NGC 4565 is viewed edge-on, creating a delicate spindle shape that inspires its nickname the "Needle Galaxy." The core of the galaxy bulges from the thin disk of its spiral arms. The core appears divided by thin obscuring lanes of dust in its galactic plane, which is typical of large spiral galaxies, including our own Milky Way. In fact, the Milky Way would resemble NGC 4565, if we were able to see it at a distance from the side. NGC 4565 resides 31 million light-years away, and spreads over 100,000 light-years in diameter. Although overlooked by the Messier catalog, it has been included as entry 38 in the Caldwell catalog (C38).

Imaging. The Needle Galaxy is surprisingly large and bright for a non-Messier galaxy, and therefore is easier to image than many other galaxies in the Best Targets. You can frame the Needle on its own, if your mount and seeing allow high resolution. A slightly larger field of view of about 30 × 20 arcmin would allow inclusion of a 2.5 × 1 arcmin galaxy, NGC 4562, which resides 13 arcmin to the southwest of the Needle. LRGB methods allow collection of the most detail in the least time, although the galaxy is bright enough for imaging with a single-shot color camera or with routine RGB methods.

Processing. If you have a rich luminance channel, try deconvolution as a first step toward sharpening. After combining your color channels and adjusting your histogram, apply additional sharpening to the dark lane of the galaxy. Balance your color, and then enrich your color to emphasize the dusty brown dark lane, the pale yellow core, and the faint blue tint in the outer arms (Fig. 5.6).

Fig. 5.6. Needle Galaxy NGC 4565. East is up.

May: Diversity of Galaxy Shapes

Camera	ST10XME (luminance), ST2000XM (RGB)
Telescopes	12-in. Meade LX200R at f/7 (luminance)
	5.5-in. TEC refractor at f/7 (RGB)
Field of view	23 × 15 arcmin
Exposures	Luminance clear 36 × 5 min, unbinned
	R 5 × 10 min, G 3 × 10 min, B 4 × 10 min, unbinned
Scale	0.6 arcsec/pixel
Limiting magnitude	6.0

May 17: Sombrero Galaxy

Designation	NGC 4594
Other names	Messier 104
Right ascension	12 h 40.0 min
Declination	−11° 42′
Magnitude	8.0
Size	9 × 4 arcmin
Constellation	Virgo

The unusual appearance of the Sombrero Galaxy M104 relates to the tilt of this spiral galaxy, only 6° from perfectly sideways. A thick band of obscuring dust along the equator of the galaxy contributes to the illusion of a Mexican hat, well seen with telescopes of 6-in. aperture or larger. The Sombrero lies 50 million light-years away in the constellation Virgo.

Imaging. Despite its listed magnitude of 8.0, the Sombrero is more difficult to image than many dimmer galaxies. First, from northern latitudes, the Sombrero is low in the sky, increasing atmospheric extinction, refraction, and turbulence. Second, the galaxy is quite small, benefiting from high resolution which places greater demands on tracking, focusing, and seeing. Third, the galaxy is pale, and this lack of color draws the eye to spatial detail, making any blurring more obvious. Finally, aside from the dark lane, the fine details are relatively low contrast compared to other edge-on spirals like the Needle and Outer Limits Galaxies. If you choose LRGB methods, acquire your luminance only at the highest altitude (near the meridian) to reduce blurring from atmospheric refraction, and thus you may have fewer luminance exposures than usual. If so, acquire your color channels unbinned to allow them to do double duty for luminance detail.

Processing. If your luminance has less imaging time than usual, you can apply your luminance with less than 100% opacity. Or, you can create a synthetic luminance by combining your RGB channels as monochrome channel. Then you can either process the synthetic luminance separately and apply it as a layer in Photoshop, or combine the synthetic luminance with the clear luminance to create a stronger luminance. My image applied the clear luminance at 80% opacity (Fig. 5.7).

May: Diversity of Galaxy Shapes

Fig. 5.7. Sombrero Galaxy M104. East is up.

Camera	ST10XME
Telescope	12-in. Meade LX200R at $f/7$
Field of view	23 × 15 arcmin
Exposures	Luminance clear 6 × 5 min, unbinned
	R, G, and B each 6 × 5 min, unbinned
Scale	0.6 arcsec/pixel
Limiting magnitude	6.0

May: Diversity of Galaxy Shapes

May 17: Whale Galaxy and Hockey Stick Galaxy

Designation	NGC 4631	NGC 4656	NGC 4627
Other names	Caldwell 32	Hidden Treasure 67	
Right ascension	12 h 42.1 min	12 h 44.0 min	12 h 42.0 min
Declination	+32° 32'	+32° 10'	+32° 34'
Magnitude	9.3	10.2	12.0
Size	15 × 3 arcmin	15 × 3 arcmin	2 × 1 arcmin
Constellation	Canes Venatici	Canes Venatici	Canes Venatici

On the right side of Fig. 5.8a, the Whale Galaxy (NGC 4631) is a large spiral galaxy, similar in size to our own Milky Way, but seen edge-on. A smaller elliptical galaxy, NGC 4627, appears to float on the back of the Whale. On the left side of the image lies the hockey-stick-shaped NGC 4656, distorted by gravitational interaction with the Whale. The close-up view in Fig. 5.8b shows that this interaction both distorts the Whale and simulates starbirth in the form of young blue clusters. All three galaxies lie 25 million light-years away.

Imaging. You can choose to frame both the Whale and Hockey Stick galaxies together with a field of view of about 50 × 35 arcmin, or you can attempt high-resolution imaging of the Whale alone if your conditions and tracking permit.

Processing. Follow processing tips for other galaxy images.

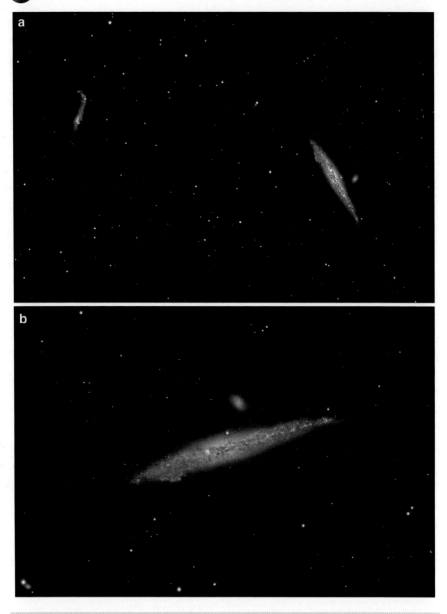

Fig. 5.8. (**a**) Hockey Stick Galaxy (*left*) and Whale Galaxy (*right*). Northeast is up. (**b**) Whale Galaxy. North is up.

May: Diversity of Galaxy Shapes

Camera (both objects)	ST10XME
Telescope	5.5-in. TEC refractor at f/7
Field of view	52×35 arcmin cropped to 44×32 arcmin
Exposures	Luminance H-alpha 24×5 min, unbinned R, G, and B each 6×5 min, binned 2×2
Scale	1.4 arcsec/pixel
Limiting magnitude	3.5

Cameras (Whale)	ST10XME (luminance), ST2000XM (RGB)
Telescopes	12-in. Meade LX200R at f/7 (luminance) 5.5-in. TEC refractor at f/7 (RGB)
Field of view	23×15 arcmin
Exposures	Luminance clear 39×5 min, unbinned R 9×10 min, G and B each 6×10 min, unbinned
Scale	0.6 arcsec/pixel
Limiting magnitude	6.0

May 19: Ringed Galaxy NGC 4725

Designation	NGC 4725
Other names	Hidden Treasure 69
Right ascension	12 h 50.4 min
Declination	+25° 30′
Magnitude	9.2
Size	11 × 8 arcmin
Constellation	Coma Berenices

Galaxy NGC 4725 at first glance is just another spiral galaxy. Yet closer inspection shows broad stretching of the core, not quite into a barred spiral, yet not a traditional spiral either. Furthermore, the central arms do not spiral, but form a ring. NGC 4725 has one of the most perfect rings of any galaxy, appearing oval to our view because of a 43° tilt from our line of sight. Active star formation in the ring appears blue-white from bright clusters of young hot stars. Beyond the ring, two faint concentric arms spiral outward. NGC 4725 is one of the brightest galaxies omitted from both the Messier and Caldwell catalogs, shining at magnitude 9.2 at a distance of 42 million light-years.

Imaging. As with many of the galaxies this month, a small field of view frames this target best. Long imaging times help to detect the faint outer arms. Luminance layering can help acquire rich imaging data in less time. If you are using luminance layering, obtain your luminance when skies are steadiest. If your site has mediocre seeing, consider binning your exposures to collect light faster.

Processing. If your data is rich, consider applying deconvolution after you have combined your luminance channel. After color combining and adjusting your histogram, apply more sharpening to the bright inner ring of the spiral arms. Avoid sharpening the foreground stars projecting over this ring. Apply a gentle blur to the faint outer arms to reduce noise. If the tenth magnitude star to the northwest of NGC 4725 appears bloated, use star-shrinking methods to correct this artifact (Fig. 5.9).

May: Diversity of Galaxy Shapes

Fig. 5.9. Ringed Galaxy NGC 4725. North is up.

Cameras	ST10XME (luminance), ST2000XM (RGB)
Telescopes	12-in. Meade LX200R at f/7 (luminance)
	5.5-in. TEC refractor at f/7 (RGB)
Field of view	22 × 15 arcmin
Exposures	Luminance clear 45 × 5 min, unbinned
	R 5 × 10 min, G 7 × 10 min, B 9 × 10 min, unbinned
Scale	0.6 arcsec/pixel
Limiting magnitude	6.0

May 19: Spiral Galaxy M94

Designation	NGC 4736
Other names	Messier 94
Right ascension	12 h 50.9 min
Declination	+41° 07′
Magnitude	8.2
Size	11 × 9 arcmin
Constellation	Canes Venatici

M94 is a bright, compact spiral galaxy with four layers of structure. The smallest of these is a central bar. Surrounding the bar are densely wound arms, with very rapid starbirth in the inner regions of the spirals. The third layer includes the outer arms, dominated by yellow and red, older suns, where earth-like environments are more likely. The fourth layer is the extremely faint outer halo of stars bound more loosely to the galaxy. M94 resides 15 million light-years away in the constellation Canes Venatici. It is much smaller than the Milky Way, spanning a width of only 30,000 light-years.

Imaging. M94 has a bright component of magnitude 8.2 measuring 6 × 4 arcmin that is relatively easy to image. Detecting the extremely faint outer halo, that extends out to 12 × 10 arcmin in this image, requires a dark sky site and long exposures. A slightly larger field of view than my image might better define this faint extension. Luminance layering helps collect light faster, but the brighter central area should be accessible to single-shot color cameras. Good flat fields are essential for M94 to distinguish the outer halo from optical vignetting.

Processing. Techniques recommended for other galaxies earlier this month also work well for M94. The critical steps for M94 are in histogram adjustments, using either digital development or curves/levels, to reveal the faint outer halo without producing excessive lightening in the brighter arms. Even if you use digital development, apply additional curves in several gentle steps to show the halo. Do not be alarmed by some noise appearing in the dim regions. Use selective smoothing methods like Gaussian Blur or "reduce noise" in these areas. Be careful if using gradient suppression tools to prevent erasure of the outer halo! (Fig. 5.10)

Fig. 5.10. Spiral Galaxy M94. East is up.

May: Diversity of Galaxy Shapes

Camera	ST10XME
Telescope	12-in. Meade LX200R at f/7
Field of view	23 × 15 arcmin cropped to 19 × 14 arcmin
Exposures	Luminance clear 13 × 5 min, unbinned R and G each 4 × 5 min, B 6 × 5 min, unbinned
Scale	0.6 arcsec/pixel
Limiting magnitude	6.0

May 21: Black Eye Galaxy M64

Designation	NGC 4826
Other names	Messier 64
Right ascension	12 h 56.7 min
Declination	+21° 41'
Magnitude	8.5
Size	9 × 5 arcmin
Constellation	Coma Berenices

The Black Eye Galaxy, M64, resides at a distance of 19 million light-years within the constellation Coma Berenices. Millions of years ago, a small galaxy tore through the center of a larger galaxy, disintegrating as it became absorbed. This interaction left the inner part of the galaxy rotating in the opposite direction from the outer regions of the galaxy. The debris of this colossal collision left a massive dust lane dominating the edge of the galactic nucleus, spurring new star formation. This dust lane is the dominant feature of M64 that is visible through small amateur telescopes, earning its name as the Black Eye!

Imaging. Among small galaxies, the Black-Eye Galaxy is relatively easy to image. Although the image shown used over 8 h of exposures from a dark sky site, I have also obtained a nice image of this galaxy from a light-polluted suburb with an 8-in. Schmidt–Cassegrain in about 3 h of exposures. A small field of view frames this object best. Longer total exposures will help extract more detail in the low-contrast spiral arms.

Processing. The only crisp details in this galaxy are in the central dust lane. Try to reveal this feature with deconvolution early in processing, and other routine sharpening tools later. Use a small radius for sharpening small details in the dust lane. The spiral arms have intrinsic low contrast, and benefit more by selective sharpening with a larger radius of 10 pixels or more (Fig. 5.11).

Cameras	ST10XME (luminance), ST2000XM (RGB)
Telescopes	12-in. Meade LX200R at f/7 (luminance)
	5.5-in. TEC refractor at f/7 (RGB)
Field of view	22 × 15 arcmin
Exposures	Luminance clear 57 × 5 min, unbinned
	R 9 × 10 min, G 6 × 10 min, B 8 × 10 min, unbinned
Scale	0.6 arcsec/pixel
Limiting magnitude	6.0

May: Diversity of Galaxy Shapes

Fig. 5.11. Black Eye Galaxy M64. West is up.

May 24: Sunflower Galaxy M63

Designation	NGC 5055
Other names	Messier 63
Right ascension	13 h 15.8 min
Declination	+42° 02'
Magnitude	8.6
Size	12 × 8 arcmin
Constellation	Canes Venatici

A welcome beacon of impending summer, the arrival of the Sunflower Galaxy in the skies of late spring herald the arrival of a warmer season. This galaxy has an unusually high number of spiral arms, tightly coiled around a bright core, giving rise to the "sunflower" name. Blue regions are illuminated by bright young stars, compared with emission nebulae glowing red from hot ionized hydrogen gas. Dense lanes of interstellar dust help to define the individual spiral arms.

With a mass of 10 billion suns and a diameter of about 60,000 light-years, the Sunflower Galaxy is only a fraction of the size of our Milky Way. The Sunflower Galaxy is closer to the size of the Pinwheel Galaxy, M33, but lies over ten times farther away. Number 63 in Messier's catalog, the Sunflower Galaxy lies at a distance of 35 million light-years in the direction of the constellation Canes Venatici.

Imaging. Frame the Sunflower galaxy with a small field of view to obtain high resolution. Rich detail extends from the central core to the outer arms. Wait for a night of good seeing to have your best chance at revealing this detail. As with most galaxies this month, luminance layering is suggested. However, the galaxy is bright enough for success with RGB technique or single-shot color, if you are willing to devote more time for imaging this target.

Processing. Follow the processing suggestions for other galaxies this month. Consider converting your RGB image to the Lab color model to enhance contrast between the yellow core and the blue outer arms. Boost color further with either saturation or match color methods to enhance the red H-II regions. If the ninth magnitude star on the west border of M63 bloats, use star-shrinking techniques (Fig. 5.12).

May: Diversity of Galaxy Shapes

Fig. 5.12. Sunflower Galaxy M63. East–northeast is up.

Cameras	ST10XME (luminance), ST2000XM (RGB)
Telescopes	12-in. Meade LX200R at f/7 (luminance)
	5.5-in. TEC refractor at f/7 (RGB)
Field of view	22 × 15 arcmin
Exposures	Luminance clear 60 × 5 min, unbinned
	R 14 × 10 min, G 9 × 10 min,
	B 10 × 10 min, unbinned
Scale	0.6 arcsec/pixel
Limiting magnitude	6.0

May 29: Whirlpool Galaxy M51

Designation	NGC 5194	NGC 5195
Other names	Messier 51	
Right ascension	13 h 29.9 min	13 h 30.0 min
Declination	+47° 12'	+47° 16'
Magnitude	8.4	10.2
Size	11 × 8 arcmin	6 × 5 arcmin
Constellation	Canes Venatici	Canes Venatici

The collision between the Whirlpool Galaxy and its companion galaxy is tearing apart the structure of the smaller galaxy, strewing a stream of stars outwards. Gravitational interaction between the two galaxies is also stretching one of the spiral arms of the Whirlpool. This interaction is triggering a burst of star formation in the Whirlpool, creating a plethora of young blue star clusters and red glowing H-II regions. Both galaxies lie 15 million light-years away.

Imaging. Large, bright, colorful, and dynamic features render the Whirlpool Galaxy among the most popular targets for astrophotographers. Rich details can be achieved with relatively short imaging times, yet longer exposures are even more rewarding. Frame the Whirlpool with a small field of view to create high resolution. Steady skies make the process much easier. My image required only 3 h of total exposures because of very calm skies that night (2-arcsec seeing). Pleasing results can also be obtained from light-polluted suburbs. LRGB methods can provide the greatest efficiency for imaging, but routine RGB methods or single-shot color can yield excellent results with longer imaging times.

Processing. If you have obtained several hours of imaging data, do not be afraid to push your techniques on this object. Begin sharpening with deconvolution. Adjust your histogram aggressively to keep the core sharp while revealing the trailing stream of stars extending from NGC 5195. Enhance your color to show the rich yellow and brown tones in NGC 5195, and the red H-II regions and blue star clusters in the Whirlpool's spiral arms. Finish up with sharpening using a small radius for small features and a larger radius to emphasize contrast between the spiral arms (Fig. 5.13).

Fig. 5.13. Whirlpool Galaxy M51. North is up.

May: Diversity of Galaxy Shapes

Camera	ST10XME
Telescope	12-in. Meade LX200R at f/7
Field of view	23 × 15 arcmin
Exposures	Luminance 18 × 5, unbinned
	R, G, and B each 6 × 5, binned 2 × 2
Scale	0.6 arcsec/pixel
Limiting magnitude	6.0

CHAPTER SIX

June: Globular Clusters and More Galaxies

June 1: Globular Cluster M3

Designation	NGC 5272
Other names	Messier 3
Right ascension	13 h 42.2 min
Declination	+28° 23'
Magnitude	5.9
Size	18×18 arcmin
Constellation	Canes Venatici

The Globular Cluster M3 lies in the constellation Canes Venatici, on the border with Bootes. Although it may contain as many stars as the Great Hercules Cluster M13, it is one-third farther away at a distance of about 33,000 light-years, and therefore somewhat dimmer and visually smaller. Furthermore, the cluster contains a dense core, with half of its 500,000 stars contained within an 11 light-year radius of its core. For comparison, only a dozen stars reside within 11 light-years of our sun. M3 is also notable for the unusually high number of variable stars, which have been used to calculate the distance to the cluster.

Imaging. On a steady night, you can image M3 using a small field of view of about 20 arcmin and a resolution under 1 arcsec/pixel to capture vivid detail in the core. On a less than perfect night, you can image successfully with a larger field of view of 40–50 arcmin and still have an attractive image. Globular cluster M3 is bright enough for routine RGB or single-shot color.

Processing. Sharpening methods are often easier to apply to just a luminance channel to avoid the creation of color noise. If you have only color data, one solution is to create a synthetic luminance. A synthetic luminance is formed by the combination of all of the red, green, and blue exposures as if they were monochrome exposures. This new channel has a higher signal-to-noise ratio than any of the individual color channels, and thus can tolerate more aggressive histogram adjustments and sharpening without excessive noise. It is then applied as a layer in Photoshop just like a regular luminance channel. For objects imaged at low altitudes subject to atmospheric refraction, or when imaging with a semiapochromatic refractor, a synthetic luminance may be sharper than a conventional luminance taken with a clear filter, if you refocus with every filter change (Fig. 6.1).

June: Globular Clusters and More Galaxies

Fig. 6.1. Globular Cluster M3. West is up.

Camera	ST10XME
Telescope	12-in. Meade LX200R at $f/7$
Field of view	23 × 15 arcmin
Exposures	R 5 × 5 min, G and B each 4 × 5 min, unbinned
Scale	0.6 arcsec/pixel
Limiting magnitude	6.0

June 6: Pinwheel Galaxy M101

Designations	NGC 5457
Other names	Messier 101
Right ascension	14 h 03.5 min
Declination	+54° 21'
Magnitude	7.9
Size	29 × 27 arcmin
Constellation	Ursa Major

The Pinwheel Galaxy is large but faint, visible as only a vague smudge through small telescopes from the suburbs. With larger telescopes from a dark rural site, one can detect the hint of spiral structure. Only photography displays its grandeur. At a distance of 27 million light-years, M101's huge size of 170,000 light-years is 70% larger than our own Milky Way Galaxy, making it one of the largest known disk galaxies. With a long-exposure astrophotograph such as this, striking asymmetry of the extended spiral arms is revealed. Abundant H-II regions populate the spiral arms.

Imaging. The Pinwheel Galaxy is rich with detail. You can try for higher resolution with a tighter field of 40 × 30 arcmin. Or, you can go for a wider field of 65 × 50 arcmin to include NGC 5474, a 5-arcmin spiral galaxy, magnitude 10.6, which resides about 44 arcmin to the south–southeast of the Pinwheel. Both approaches should yield pleasing results. Either RGB methods or single-shot color will work fine, but more faint detail can be extracted with LRGB. If you want to emphasize the H-II regions, acquire part of your luminance with either a red or an H-alpha filter.

Processing. Begin processing the Pinwheel using routine RGB or LRGB methods. If you have obtained some filtered luminance exposures, consider blending the clear and filtered luminance as a new combined luminance before applying in Photoshop. You may also want to strengthen your red channel with any H-alpha exposures before color combining, to provide a better match to your filtered luminance. As with most bright galaxies, sharpen the bright core and enrich color to reveal the blue arms, red H-II regions, and yellow core (Fig. 6.2).

Fig. 6.2. Pinwheel Galaxy M101. North–northwest is up.

June: Globular Clusters and More Galaxies

Camera	ST10XME
Telescope	5.5-in. TEC refractor at f/7
Field of view	52 × 34 arcmin
Exposures	Luminance clear 23 × 5 min, Ha 7 × 10, unbinned
	R 4 × 5 min, G and B each 3 × 5 min, binned 2 × 2
Scale	1.4 arcsec/pixel
Limiting magnitude	6.0

June 25: Splinter Galaxy

Designation	NGC 5907
Other names	
Right ascension	15 h 15.9 min
Declination	+56° 20'
Magnitude	10.4
Size	12 × 2 arcmin
Constellation	Draco

The Splinter Galaxy (NGC 5907) is an edge-on spiral galaxy at a distance of 40 million light-years. Dust lanes partially obscure the disk of hundreds of millions of stars. Slight warping of the edges of the galaxy is thought to be related to interactions with two nearby dwarf galaxies, which also contribute to a faint halo ring that is too dim to see in my image.

Imaging. The Splinter Galaxy is simple and elegant, but requires high resolution of 1 arcsec/pixel or smaller to show detail. A steady night with good seeing is needed to get any resolution of the dark lane. Good tracking and accurate focus are also essential for sharp detail. LRGB methods are ideal because the Splinter is relatively faint. If you have two CCD cameras, you can use hybrid imaging to obtain your luminance with one camera and your RGB channels with the other, to obtain all of your imaging data in less time.

Processing. After combining your exposures into channels, consider applying deconvolution to your luminance to start sharpening details. After combining color and balancing color in your RGB channels, enhance the color to emphasize the contrast between the yellow galaxy and several bright blue-white foreground stars. This can be achieved by either increasing saturation or match color adjustments. Another option is to convert your image to a Lab color model, as discussed in the section "Color Enrichment" in Chap. 15. Sharpen the dark lane in the galaxy as much as noise allows. Finally, apply some blurring (or Photoshop's reduce noise filter) to the background to smooth away noise and grainy color (Fig. 6.3).

June: Globular Clusters and More Galaxies

Fig. 6.3. Splinter Galaxy. North is up.

Camera	ST10XME (luminance), ST2000XM (RGB)
Telescopes	12-in. Meade LX200R at f/7 (luminance)
	5.5-in. TEC refractor at f/7 (RGB)
Field of view	23 × 15 arcmin
Exposures	Luminance clear 36 × 5 min, unbinned
	R, G, and B each 6 × 10 min, unbinned
Scale	0.6 arcsec/pixel
Limiting magnitude	6.0

June 26: Globular Cluster M5

Designation	NGC 5904
Other names	Messier 5
Right ascension	15 h 18.5 min
Declination	+02° 05′
Magnitude	5.7
Size	23 × 23 arcmin
Constellation	Serpens

The Globular Cluster M5 resides 24,500-light years away in the constellation Serpens. At an age of 13 billion years, it is one of the most ancient globular clusters known, forming only a few billion years after the Big Bang. M5 is one of the 160 globular clusters known to reside in a spherical halo around the Milky Way's galactic center. M5 is one of the larger globular clusters, containing about 100,000 stars within a diameter of 165 light-years.

Imaging. This image shows one of the options for framing the large globular clusters like M5. A moderate size field of view of 30–50 arcmin reinforces the concentration of stars within the globular cluster. The other option, as shown in the next two targets (M13 and M12), is to frame the cluster more tightly to resolve more of the core. Either approach can be successful, but the larger field of view places less demands on seeing, tracking, guiding, and focusing. Decrease your exposure times as needed to prevent overexposing of the core. Although I used LRGB methods for this image, routine RGB methods or single-shot color can work just as well for the brighter globular clusters.

Processing. For globular clusters, your goal is to resolve the center of the cluster while revealing the faint outer edges of the cluster. If deconvolution creates a weird mosaic pattern to the core of the cluster, then reserve sharpening to later stages of processing. For histogram adjustments of globular clusters, curves and levels are better than digital development. The star with a magnitude of 5.0 to the southwest helps to balance the image, but if it appears excessively bloated or otherwise detracting, apply star-shrinking methods (see section "Final Cleanup" in Chap. 15) (Fig. 6.4).

Fig. 6.4. Globular Cluster M5. North is up.

June: Globular Clusters and More Galaxies

Camera	ST10XME
Telescope	5.5-in. TEC refractor at $f/7$
Field of view	52 × 35 arcmin
Exposures	Luminance clear 10 × 3 min, unbinned R 5 × 3 min, G 3 × 3 min, B 4 × 3 min, binned 2 × 2
Scale	1.4 arcsec/pixel
Limiting magnitude	6.0

CHAPTER SEVEN

July: Just Globular Clusters

July 17: Great Hercules Cluster

Designations	NGC 6205
Other names	Messier 13
Right ascension	16 h 41.7 min
Declination	+36° 28'
Magnitude	5.7
Size	20 × 20 arcmin
Constellation	Hercules

The Great Hercules Cluster M13 is the colossus of globular clusters visible in the northern hemisphere. M13 can be seen with the unaided eye from a dark sky site. It packs about a half million stars in a region only 150 light-years across, orbiting our galaxy like a giant satellite 24,000 light-years away. The brightest stars in this image are red giants, thousands of times brighter than our sun. At its core, the glare of brilliant suns would obscure the remainder of the universe. Of course, planets are unlikely in the center of a globular cluster, as frequent gravitational interactions with other stars would strip away planets.

Imaging. This image of M13 is framed tightly to create high resolution. Innumerable small stars are resolved around the edge of the cluster. Such high-resolution imaging, at less than 1 arcsec/pixel, requires steady skies, accurate focusing, good tracking, and autoguiding. If your skies or mount are not up to the task, then consider a larger field of view as in the image of M5 shown in section "June 26: Globular Cluster M5" in Chap. 6. The bright globular clusters can be imaged successfully with RGB or single-shot color methods. Luminance layering is usually not necessary for globulars.

Processing. The best images of globular clusters show color matched precisely to individual resolved stars. This requires precise alignment of your color channels. Inspect your channels after alignment to confirm proper alignment. Although it is best to complete alignment before color combine, a last minute fix in Photoshop is possible. To do this on a single RGB image, go to channels, choose just the color channel that is fringing, click RGB to show all colors, then select all and use either the move tool or transform functions to shift the offending channel (Fig. 7.1).

July: Just Globular Clusters

Fig. 7.1. Hercules Cluster M13. North is up.

Camera	ST10XME
Telescope	12-in. Meade LX200R at *f*/7
Field of view	23 × 15 arcmin
Exposures	Luminance clear 15 × 3 min, unbinned R and G each 4 × 3 min, B 6 × 3 min, binned 2 × 2
Scale	0.6 arcsec/pixel
Limiting magnitude	6.0

July 18: Globular Cluster M12

Designation	NGC 6218
Other names	Messier 12
Right ascension	16 h 47.2 min
Declination	–01° 57′
Magnitude	6.1
Size	16 × 16 arcmin
Constellation	Ophiuchus

The Globular Cluster M12 is the brightest and most northern of seven globular clusters originally cataloged by Charles Messier over 200 years ago in the constellation Ophiuchus. At a distance of 16,000 light-years, M12 shows less concentration of stars than most other globular clusters. Many of its original low-mass stars may have been stripped away by gravitational interactions with the Milky Way.

The scientific study of globular clusters provides important clues to the age of the universe and the formation of our galaxy. The age of globular clusters can be determined by spectral analysis, which shows the content of metals (metallicity). Older stars, forming in the early years of the universe, have less heavy elements because fewer supernovas had occurred to distribute these elements to the interstellar void. Globular clusters are among the oldest objects whose age can be measured, giving a lower constraint on the age of the universe. Furthermore, the spatial distribution of globulars around our galaxy, combined with knowledge of their age, helps to define factors influencing the formation of our galaxy.

Imaging. As with other globular clusters, you can frame M12 with either a medium size field of view or with a smaller field of view for higher resolution. Either LRGB, RGB, or single-shot color can be successful with the larger globulars.

Processing. As with other globular clusters, first calibrate images, then align and combine your channels, and consider applying deconvolution to begin sharpening. Then color combine, and adjust your histogram with curves and levels. Use sharpening aggressively, and boost color with routine enhancement methods (Fig. 7.2).

Fig. 7.2. Globular Cluster M12. North is up.

July: Just Globular Clusters

Camera	ST10XME (luminance), ST2000XM (RGB)
Telescopes	12-in. Meade LX200R at f/7 (luminance) 5.5-in. TEC refractor at f/7 (RGB)
Field of view	23 × 15 arcmin
Exposures	Luminance clear 13 × 5 min, unbinned R 11 × 3 min, G and B each 10 × 3 min, unbinned
Scale	0.6 arcsec/pixel
Limiting magnitude	6.0

CHAPTER EIGHT

August: Planetary and Emission Nebulae

August 5: Cat's Eye Nebula

Designation	NGC 6543
Other names	Caldwell 6
Right ascension	17 h 58.6 min
Declination	+66° 38'
Magnitude	8.1
Size	0.3/5.8 arcmin
Constellation	Draco

The brilliant teal glow in the center of the Cat's Eye Nebula impresses visual astronomers with large telescopes. This central bright region, 20 arcsec in diameter, is just a small portion of the much larger (386 arcsec) outer halo that is a 1,000-fold dimmer, and only appears in long-exposure images such as this one. Many fainter intermediate layers each reflect episodic rigors of a dying star. The Cat's Eye Nebula resides at a distance of 3,000 light-years.

Imaging. As with other small planetary nebulae, image the Cat's Eye with the highest resolution that your skies, mount, and tracking will allow. The inner nebula is bright and easy to capture even with a single-shot color camera or routine RGB imaging. Be careful not to overexpose the central region. On the other hand, the outer faint layers are better revealed with a clear filter using long exposures that intentionally overexpose the center. The overexposed center can later be subtracted during processing. Furthermore, because the outer layers are lower in both contrast and detail, you can acquire the clear exposures binned 2 × 2 to improve your signal-to-noise ratio in the dim areas.

Processing. Calibrate and align your images routinely. If you want to try deconvolution to get a head start on sharpening the central area, then only select your very sharpest unbinned RGB individual exposures to combine for a synthetic luminance channel. Combine your RGB channels routinely. If you have an overexposed clear luminance for the outer nebula, create several copies each with slightly more aggressive digital development adjustments. Apply these as multiple luminance layers, each with 10–20% opacity; for each, select the central core, feather generously, and cut out the area that was burnt out by digital development (Fig. 8.1).

August: Planetary and Emission Nebulae

Fig. 8.1. Cat's Eye Planetary Nebula. North is up.

Camera	ST10XME
Telescope	12-in. Meade LX200R at *f*/7
Field of view	22 × 14 arcmin
Exposures	Luminance clear 63 × 4 min, unbinned R and B each 11 × 4 min, G 7 × 4 min, unbinned
Scale	0.6 arcsec/pixel
Limiting magnitude	6.0

August 6–7: Trifid and Lagoon Nebulae

Designations	NGC 6514	NGC 6523
Other names	Messier 20, Trifid	Messier 8, Lagoon
Right ascension	18 h 02.4 min	18 h 03.7 min
Declination	−23° 02'	−24° 23'
Magnitude	6.3	3.6
Size	29 × 27 arcmin	90 × 40 arcmin
Constellation	Sagittarius	Sagittarius

The Trifid Nebula was named by John Herschel over 200 years ago to describe its trilobed appearance. A bright young triple star at the center of the Trifid excites and illuminates the surrounding hydrogen clouds. The three components of the triple star are too close to each other to be distinguished on this image. Above the emission nebula, a faint blue reflection nebula is seen.

A wide field view of 2° shows both the Trifid Nebula and the larger Lagoon Nebula. Both the Trifid and Lagoon nebulae are parts of the same cloud of gas harboring star formation, 5,200 light-years away.

Imaging. The Trifid alone can be framed with a medium size field of view of 30–40 arcmin. A wider field of 120 arcmin or more is needed to frame both the Trifid and Lagoon together. Both are bright targets that are captured well with either RGB methods or single-shot color. If you image these southern objects from northern latitudes, RGB may be better than LRGB because atmospheric refraction will blur luminance images obtained at low altitudes. Obtain your blue exposures near the meridian to reduce atmospheric extinction.

Processing. Routine processing methods work well for the Trifid and Lagoon (Fig. 8.2).

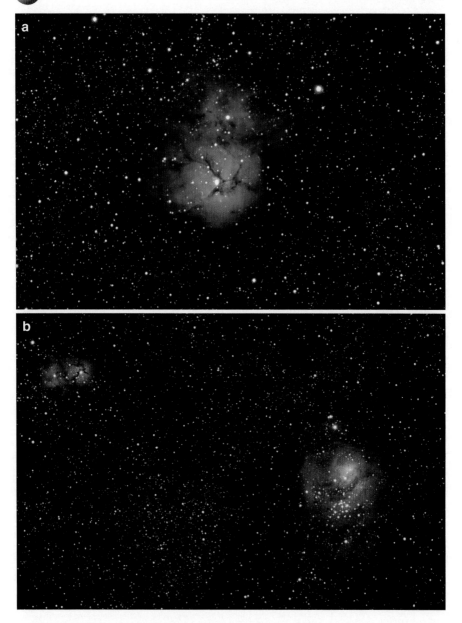

Fig. 8.2. (**a**) Trifid Nebula M20. North–northeast is up. (**b**) Trifid Nebula (*left*) and Lagoon Nebula (*right*). West is up.

August: Planetary and Emission Nebulae

Camera (both images)	ST10XME
Telescope (Trifid)	5.5-in. TEC refractor at f/7
Field of view	52 × 35 arcmin
Exposures	Luminance IDAS 3 × 5 min, unbinned R and G each 1 × 5 min, B 2 × 5 min, binned 2 × 2
Scale	1.4 arcsec/pixel
Telescope (M8 and M20)	3.5-in. Takahashi refractor at f/4.5
Field of view	126 × 85 arcmin
Exposures	R, G, and B each 11 × 3 min, unbinned
Scale	3.5 arcsec/pixel
Limiting magnitude	6.0

August 7: Lagoon Nebula

Designations	NGC 6523
Other names	Messier 8
Right ascension	18 h 03.7 min
Declination	−24° 23'
Magnitude	3.6
Size	90 × 40 arcmin
Constellation	Sagittarius

This close-up view of the Lagoon Nebula shows its central dark lane of obscuring dust, appearing like a lagoon when examined visually through a telescope. Other more focal dark areas within the Lagoon Nebula are collapsing protostellar clouds called Bok Globules. The brightest region in the nebula, called the hourglass nebula, is illuminated by two supergiant stars. The rest of the nebula is energized by the young open cluster NGC 6530 emerging in the other side of the dark "lagoon." The center of the nebula has a magenta tone because dust reflects some of the light from the bright stars, combining blue reflection with red emission.

Imaging. This image of the Lagoon Nebula is framed about as tight as you should go. A slightly larger field of view might look less crowded. To capture the Trifid Nebula in the same frame, you would need to increase your field of view to 2°. The Lagoon Nebula is bright enough to image well either with routine RGB methods or with a single-shot color camera. A clear luminance would introduce blurring from atmospheric refraction due to the low altitude of this object from northern latitudes. However, the red exposures can double as a filtered luminance, which was done for this image, boosting detail of the red emission.

Processing. After image calibration, debloom the brighter stars. Align and combine with routine methods. If you plan to use your red channel as a filtered luminance, then apply deconvolution to a copy of your red channel and save this separately from the original red channel. Continue with color combine of your RGB channels. Adjust your histograms with either digital development or curves/levels, or with both. Before you apply sharpening tools to the brighter areas of the nebula, deselect the brighter stars to prevent artificial halos. When balancing color, allow the central reflection component to give a magenta hue (Fig. 8.3).

Fig. 8.3. Lagoon Nebula. West is up.

Camera   ST10XME
Telescope   5.5-in. TEC refractor at f/7
Field of view   52 × 35 arcmin
Exposures   R 11 × 3 min, G 8 × 3 min, B 9 × 3 min, unbinned
Scale   1.4 arcsec/pixel
Limiting magnitude 6.0

August 10: Eagle Nebula

Designation	NGC 6611
Other names	Messier 16, IC 4703
Right ascension	18 h 18.8 min
Declination	−13° 47'
Magnitude	6.0
Size	35 × 28 arcmin
Constellation	Serpens

Within the Eagle Nebula (IC 4703), a billowing cloud of interstellar gas and dust has entered an active phase of star formation. Open star cluster M16 (NGC 6611) is emerging in the upper right corner of this great gaseous and dusty cloud. Ultraviolet radiation from these young massive hot stars excites the hydrogen gas in the surrounding Eagle Nebula to shine by red emission light. The nebula continues to form new stars near the dark central area of condensation, which Hubble photographers have called the "pillars of creation." The Eagle Nebula and M16 reside 7,000 light-years distant.

Imaging. Many options are possible for framing the Eagle Nebula. A wider field of 40–50 arcmin or more helps to frame the wings of the eagle. A very wide field of 2 × 3° can include both the Eagle Nebula and the Swan Nebula. A smaller field of view allows higher resolution of the "pillars" in the center of this view. Because of the low altitude of this object from northern latitudes, a clear luminance is not recommended. A filtered luminance, with either a red or an H-alpha filter, can help reveal detail in the central pillars. If you plan to obtain an H-alpha luminance, use longer exposures.

Processing. Begin processing with routine methods. If you are using an H-alpha luminance, skip deconvolution, which can contribute to artifacts when you apply your H-alpha channel to the RGB channels. H-alpha luminance is great for extracting detail from emission nebula, but can create harsh artifacts. Therefore, as discussed in the section "Luminance Layering" in Chap. 15, consider blending your H-alpha channel with your red channel both for RGB combine and for luminance. Apply aggressive sharpening to the central pillars and other areas of high contrast (Fig. 8.4).

Fig. 8.4. Eagle Nebula. North is up.

August: Planetary and Emission Nebulae

Camera	ST10XME
Telescope	12-in. Meade LX200R at $f/7$
Field of view	22 × 15 arcmin
Exposures	Luminance Ha 9 × 10 min, unbinned R 17 × 3 min, G 15 × 3 min, B 22 × 3 min, unbinned
Scale	0.6 arcsec/pixel
Limiting magnitude	6.0

August 11: Swan Nebula

Designations	NGC 6618
Other names	Messier 17, Omega, Swan
Right ascension	18 h 20.8 min
Declination	−16° 11′
Magnitude	6.0
Size	46 × 37 arcmin
Constellation	Sagittarius

The Swan Nebula M17, like many other emission nebulae, shows red light emitted from hydrogen gas clouds excited by the high energy of young stars. Unlike other nebulae of its class, the bright stars that are energizing the Swan nebula are concealed by dense dust. Some of this dust in the center of the Swan both obscures and reflects the light from these bright stars. White and blue light reflected by dust, combined with red light emitted by hydrogen, can create a mix of pink and magenta tones in the center of the Swan.

The Swan Nebula lies at the edge of a large dense molecular cloud 5,500 light-years away within the Sagittarius arm of the Milky Way.

Imaging. The brightest part of this nebula, which has been described as a swan, a duck, or the Greek letter omega, is only about 15 arcmin across and therefore can be framed with a small field of view. However, there is much more extensive faint nebulosity extending out to 50 arcmin, and so a larger field of view can also be used successfully. To capture the most detail in the H-II regions, a filtered luminance with either a red filter or an H-alpha filter is suggested.

Processing. Process the Swan Nebula similar to methods for the Eagle Nebula, as described in the section "August 10: Eagle Nebula". Adjusting color saturation is more difficult for the Swan Nebula. The central area of nebulosity is very bright and tends to lose color saturation becoming pink or white when you try to brighten the dimmer parts of the nebula. Apply curves gradually, possibly using selections or layer masks to avoid excessive brightening centrally. The Swan Nebula has inherently lower contrast than the Eagle Nebula. Therefore, when sharpening, do not be surprised if your results seem muted (Fig. 8.5).

August: Planetary and Emission Nebulae

Fig. 8.5. Swan Nebula M17. South is up.

Camera	ST10XME
Telescope	12-in. Meade LX200R at $f/7$
Field of view	22 × 15 arcmin
Exposures	Luminance Ha 6 × 10 min, unbinned R 8 × 3 min, G 6 × 3 min, B 9 × 3 min, unbinned
Scale	0.6 arcsec/pixel
Limiting magnitude	6.0

August 15: Globular Cluster M22

Designation	NGC 6656
Other names	Messier 22
Right ascension	18 h 36.4 min
Declination	−23° 54′
Magnitude	5.1
Size	32 × 32 arcmin
Constellation	Sagittarius

Although the Hercules Cluster M13 is widely recognized as the most striking globular cluster visible from northern latitudes, M22 is actually brighter but harder to see due to its low altitude. At a distance of about 10,000 light-years, its 70,000 suns are packed tightly into a diameter of under 100 light-years. As viewed from Earth, M22 spans an area of the size of the full moon, but only the central region is visible through telescopes.

Imaging. Globular Cluster M22 is quite large, and should be framed by a field of view of at least 30 arcmin. Because of its low declination, M22 will be at a low altitude from northern latitudes. The low altitude increases the impact of atmospheric seeing, so you may prefer a larger field of view to create a larger pixel size, which reduces the effect of seeing on your image. As with other bright globulars, luminance layering is not needed, and in fact can blur your image due to atmospheric refraction for objects at low altitudes. For RGB imaging, obtain your blue channel near the meridian to suppress atmospheric extinction. If you are using single-shot color, try to obtain all of your images above 20° altitude.

Processing. As with other globular clusters, make sure your color channels are aligned well before color combine. Consider creating a synthetic luminance, as discussed with processing of M3 (June 1). Deconvolution sometimes causes a strange mosaic pattern in the core of globulars; if this occurs to your image, postpone sharpening to a later step. Adjust your histogram with curves and levels using multiple gradual steps to reveal the outer regions of the cluster without burning out the core. After balancing the color, boost the color only in the bright stars using either the color range selection set to highlights or the magic wand tool to select the bright stars, feathering your selection by 1 or 2 pixels (Fig. 8.6).

Fig. 8.6. Globular Cluster M22. North–northwest is up.

August: Planetary and Emission Nebulae

Camera	ST10XME
Telescope	5.5-in. TEC refractor at f/7
Field of view	52 × 35 arcmin
Exposures	R 7×5 min, G 6×5 min, B 9×5 min, unbinned
Scale	1.4 arcsec/pixel
Limiting magnitude	6.0

August 18: Wild Duck Cluster

Designation	NGC 6705
Other names	Messier 11
Right ascension	18 h 51.1 min
Declination	–06° 16'
Magnitude	5.8
Size	11 × 11 arcmin
Constellation	Scutum

The Wild Duck Cluster was first recorded in 1681 by Gottfried Kirch, and 83 years later entered into Messier's catalog as Number 11 of faint comet-like objects. It is not the easiest cluster to find, lurking 6,000 light-years away in the insignificant constellation Scutum. Yet, this object looks even more spectacular through a telescope than in this image. M11 is a rich open cluster containing almost 3,000 stars, of which 500 are bright enough to dazzle through a backyard telescope. The delta-shape orientation of the stars evokes the image of a flock of ducks, thus the name "Wild Duck Cluster."

Imaging. The Wild Duck Cluster requires a frame of at least 20 arcmin to allow the concentration of stars to be distinguished from the general background stars. The individual stars are bright enough for either single-shot color or RGB methods. Luminance layering is not needed in most cases. However, if your mount tracks poorly, you may want to try to shorten your exposures with luminance layering. A luminance exposure can capture data about four times faster than a color channel, allowing less time for stars to elongate or blur.

Processing. As with most open clusters, perfect image alignment is essential to prevent ugly asymmetric color fringing. Deconvolution is usually not needed for open clusters. After combining your color channels, adjust your histogram with curves and levels to reveal the full extent of the cluster. Avoid digital development with open clusters. Sharpen your images gently to define the central stars more clearly. After balancing your color, enhance the color gently either by boosting saturation or using Photoshop's match color adjustment. Finally, compensate for any light-pollution gradients (see section "Dealing Out Gradients" in Chap. 15) (Fig. 8.7).

August: Planetary and Emission Nebulae

Fig. 8.7. Wild Duck Cluster M11. North is up.

Camera	ST10XME
Telescope	8-in. Celestron Schmidt–Cassegrain at $f/7$
Field of view	33 × 22 arcmin
Exposures	Luminance IDAS 40 × 1 min, unbinned R and G each 12 × 1 min, B 16 × 1 min, unbinned
Scale	0.9 arcsec/pixel
Limiting magnitude	3.5

August 19: Ring Nebula

Designation	NGC 6720	IC 1296
Other names	Messier 57	
Right ascension	18 h 53.6 min	18 h 53.3 min
Declination	+33° 02'	+33° 04'
Magnitude	8.8	14.3
Size	3 × 3 arcmin	1 × 1.3 arcmin
Constellation	Lyra	Lyra

Over 200 years ago, the famous English astronomer William Herschel found the Ring Nebula M57 to resemble the planet newly discovered by him, Uranus, and so introduced the term "Planetary Nebula." These objects are created during the dying gasps of mid-sized suns, when their central supply of nuclear fuel becomes exhausted. As the star collapses into a dense hot white dwarf, the outer layers blossom in a brilliant splash of color, becoming energized and illuminated by ultraviolet radiation from the white hot ember at their center.

Whereas the Ring Nebula is about 2,000 light-years away, the barred spiral galaxy IC 1296 at the right of this image is about 200 million light-years distant!

Imaging. The Ring Nebula is best displayed with a small field of view to reveal its features. High-resolution imaging demands steady skies, accurate tracking, and precise focus. The main ring is bright, and easily captured with either RGB methods or single-shot color. The outer halo is extremely faint, but can be imaged using a variation of luminance layering. Use very long H-alpha or red exposures, binned 3 × 3, to hugely improve the signal-to-noise in the dim halo. The main ring is overexposed in this layer, but will be subtracted during processing.

Processing. First, process your RGB with routine methods. Then, process the H-alpha channel like a luminance, and scale to your RGB channels. Create several copies each with slightly more aggressive digital development. Apply these as multiple luminance layers, each with 10–20% opacity. Select, feather, and cut out the center core that was burnt out by digital development. If the outer halo lacks color, enrich the red channel with 10% of the H-alpha channel. To reveal the galaxy IC 1296, select the area of the galaxy, feather, and apply curves (Fig. 8.8).

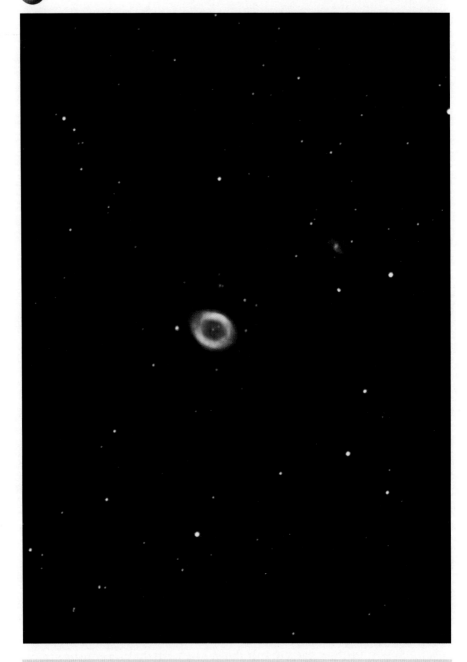

Fig. 8.8. Ring Nebula and IC 1296. North is up.

August: Planetary and Emission Nebulae

Camera	ST10XME
Telescope	12-in. Meade LX200R at $f/7$
Field of view	23 × 15 arcmin cropped to 17 × 12 arcmin
Exposures	H-alpha 6 × 15 min, binned 3 × 3 R and B each 11 × 5 min, G 10 × 5 min, unbinned
Scale	0.6 arcsec/pixel
Limiting magnitude	6.0

CHAPTER NINE

September: Autumn Assortment

September 1: Barnard's Galaxy

Designation	NGC 6822
Other names	Caldwell 57
Right ascension	19 h 44.9 min
Declination	−14° 48′
Magnitude	8.8
Size	15 × 14 arcmin
Constellation	Sagittarius

Edward Barnard first observed this object in 1884 with a 5-in. refractor. In 1924, Edwin Hubble used the 100-in. Mt. Wilson telescope to identify 11 Cepheid variables in this object. In the December 1925 issue of the Astrophysical Journal, Hubble reported that the period–luminosity relationship of these variable stars placed NGC 6822 at least 700,000 light-years distant. This proved that Barnard's discovery was in fact outside of the Milky Way, and far more distant than the Magellanic Clouds, which were the first confirmed extragalactic structures. Barnard's Galaxy is an irregular dwarf galaxy with a bar along its north–south axis. At the north border (upper right), several H-II regions can be seen glowing red. In the lower left, a band of stars are grouped into a short arm. Current data places Barnard's Galaxy about 1.8 million light-years away.

Imaging. Barnard's Galaxy is much more difficult to image than its magnitude of 8.8 and size of 15 arcmin may suggest. First, this target is low in the sky for northern observers, which both introduces more atmospheric effects and limits time for imaging. Second, its light is spread out yielding low contrast. Even with 4 h of exposures from a dark sky site, this image is somewhat disappointing, which I attribute to my latitude of 43° north. If your site is farther south, you should have a better opportunity. Imaging this object requires a dark sky site, so do not try this from the suburbs.

Processing. Barnard's Galaxy is low in contrast and therefore may not benefit from deconvolution or other sharpening methods. After balancing color, boost color saturation to emphasize the H-II regions. If you are at a northern latitude, then you may have gradients from imaging at low altitudes; apply gradient reduction methods if needed (Fig. 9.1).

September: Autumn Assortment

Fig. 9.1. Barnard's Galaxy. Northeast is up.

Camera	ST10XME
Telescope	12-in. Meade LX200R at *f*/7
Field of view	23 × 15 arcmin
Exposures	R, G, and B each 12 × 5 min, unbinned
Scale	0.6 arcsec/pixel
Limiting magnitude	6.0

September 5: Dumbbell Nebula

Designations	NGC 6853
Other names	Messier 27
Right ascension	19 h 59.6 min
Declination	+22° 43′
Magnitude	7.3
Size	7×7 arcmin
Constellation	Vulpecula

The Dumbbell Nebula M27 was the first planetary nebula seen by human eyes, discovered by Charles Messier in 1764. John Herschel later coined the Dumbbell name. Even a small telescope shows a distinct hourglass with a surrounding egg-shaped cloud of gas. The central teal color arises from the emission lines of doubly ionized oxygen at wavelengths around 500 nm. These emission lines can only occur in a low-density environment like a planetary nebula.

Imaging. Frame the Dumbbell Nebula with a small field of view between 15 and 30 arcmin, which benefits from steady skies, accurate tracking, and precise focusing. This nebula is bright and easy to capture with either a single-shot color camera or with routine RGB methods. With very long exposures, a faint outer halo can be demonstrated, which is not shown on my image. If you want to try to capture the outer halo, long exposures with an H-alpha filter can be tried with the methods described in section "August 19: Ring Nebula" in Chap. 8. The outer shell increases the overall size of the nebula to 15 arcmin, so choose a slightly larger field of view if you are trying to image the faint halo.

Processing. Begin with calibration and alignment of your exposures. Combine your color channels with routine methods. Adjust your histogram with either digital development or curves/levels to create a bright outer rim and a more subdued center. Balance color carefully to maintain a teal center and a red rim. Enrich the color with either a boost in saturation or with "match color" in Photoshop. Sharpen the brighter red clouds with either unsharp masking, high-pass filtering, or Photoshop's smart sharpen filter. Smooth the fainter teal nebulosity to suppress noise. Finally, remove any background gradients (Fig. 9.2).

Fig. 9.2. Dumbbell Nebula. North is up.

September: Autumn Assortment

Camera	ST10XME
Telescope	12-in. Meade LX200R at f/7
Field of view	23 × 15 arcmin
Exposures	R, G, and B each 12 × 5 min, unbinned
Scale	0.6 arcsec/pixel
Limiting magnitude	6.0

September 8: Crescent Nebula

Designation	NGC 6888
Other names	Caldwell 27
Right ascension	20 h 12.0 min
Declination	+38° 21'
Magnitude	8.8
Size	20 × 10 arcmin
Constellation	Cygnus

The Crescent Nebula is an expanding shell of hydrogen gas measuring 16 × 25 light-years across. The name "crescent" is evoked by its ethereal glow in the shape of a young moon, when viewed through large backyard telescopes. The Crescent Nebula is illuminated by a central brilliant Wolf–Rayet star at its center. Wolf–Rayet stars possess a powerful high-velocity solar wind that energizes the gases that shed earlier in the star's evolution. A similar process creates the spectacular Bubble Nebula and Thor's Helmet. Wolf–Rayet stars are a common phase in the evolution of huge type O stars, but are rare because of their short lifespan, typically ending in a supernova explosion within a million years.

Imaging. The Crescent is nicely framed with a field of view of about 30–40 arcmin. The more luminous rim of the Crescent is bright enough to image with a single-shot color camera or RGB methods. The delicate central tendrils are revealed with H-alpha filters and luminance layering.

Processing. Routine processing steps should be followed. If you have obtained an H-alpha channel, blend these exposures with your red channel. The blended exposures can be weighted mostly by the red channel if you plan to use the result as a stronger red channel for the RGB combine. If your goal is to modify the H-alpha luminance to better match the RGB exposures, then weight the combination with mostly H-alpha. For this image, the red channel was combined with the H-alpha exposures for a blended luminance, which was applied in Photoshop with an opacity of 70% (Fig. 9.3).

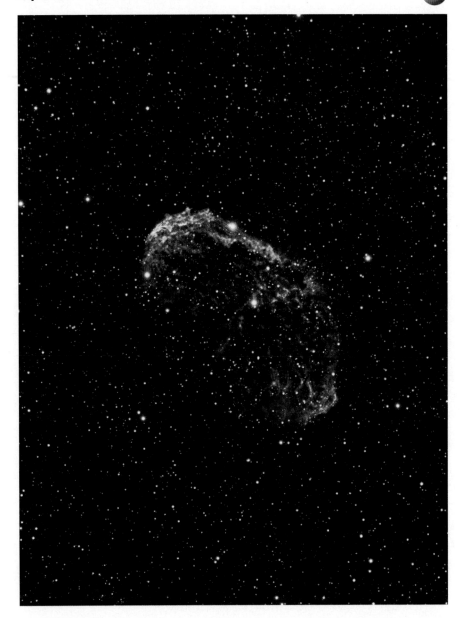

Fig. 9.3. Crescent Nebula. North is up.

Camera	ST2000XM
Telescope	5.5-in. TEC refractor at f/7
Field of view	41 × 31 arcmin
Exposures	Luminance Ha 7 × 20 min, unbinned R 9 × 10 min, G 6 × 10 min, B 8 × 10 min, unbinned
Scale	1.6 arcsec/pixel
Limiting magnitude	6.0

September 14: Fireworks Galaxy and Cluster NGC 6939

Designation	NGC 6946	NGC 6939
Other names	Caldwell 12	Melotte 231
Right ascension	20 h 34.8 min	20 h 31.5 min
Declination	+60° 09'	+60° 40'
Magnitude	8.9	7.8
Size	11 × 10 arcmin	8 × 7 arcmin
Constellation	Cepheus	Coma Berenices

Galaxy NGC 6946 resides at a distance of 19 million light-years, compared to only a few thousand light years for the rich open cluster NGC 6939.

Imaging. You can frame both the galaxy and the open cluster using a 70 × 50 arcmin field of view. If your skies and tracking allow, you can use a smaller field of view for high-resolution imaging of the Fireworks Galaxy (Fig. 9.4).

Camera	ST10XME
Telescope	4-in. Astro-Physics refractor at f/6
Field of view	82 × 55 arcmin
Exposures	Luminance IDAS 31 × 3 min, unbinned R 6 × 3 min, G 5 × 3 min, B 7 × 3 min, binned 2 × 2
Scale	2.3 arcsec/pixel
Limiting magnitude	3.5

Camera	ST10XME
Telescope	12-in. Meade LX200R at f/7
Field of view	23 × 15 arcmin cropped to 18 × 15 arcmin
Exposures	Luminance clear 38 × 5 min, unbinned R, G, and B each 12 × 5 min, binned 2 × 2
Scale	0.6 arcsec/pixel
Limiting magnitude	6.0

Fig. 9.4. (**a**) Fireworks Galaxy NGC 6946 and Open Cluster NGC 6939. Northeast is up. (**b**) Fireworks Galaxy NGC 6946. North is up.

September 17: Veil Nebula, "Witch's Broom" of Veil

Designations	NGC 6960	NGC 6992/6995
Other names	Caldwell 34, Witch's Broom	Caldwell 33, Network
Right ascension	20 h 45.7 min	20 h 56.4 min
Declination	+30° 43'	+31° 43'
Magnitude	–	–
Size	70 × 6 arcmin	60 × 8 arcmin
Constellation	Cygnus	Cygnus

About 10,000 years ago, a supergiant star in the constellation Cygnus exploded as a supernova. Despite its distance of about 2,500 light-years, the supernova may have shown as bright as the moon for weeks, but no records of its observation have been discovered. The gas expelled by this explosion has been expanding as a giant bubble, now colliding with hydrogen, oxygen, and sulfur gases that were previously shed by the dying star. This collision creates a shock front, exciting the gases similar to an emission nebula with red hydrogen and teal oxygen. (Imaging and Processing are discussed in the section "September 19: Eastern Loop of Veil: The Network Nebula".) (Fig 9.5)

Camera	ST10XME
Telescope	Nikon 180-mm camera lens at f/2.8
Field of view	4 × 3°
Exposures	R and G each 14 × 3 min, B 21 × 3 min, binned 2 × 2
Scale	15 arcsec/pixel
Limiting magnitude	3.5

Camera	ST2000XM
Telescope	5.5-in. TEC refractor at f/7
Field of view	41 × 31 arcmin
Exposures	Luminance Ha 7 × 20 min, unbinned R 6 × 10 min, G 5 × 10 min, B 8 × 10 min, unbinned
Scale	1.6 arcsec/pixel
Limiting magnitude	6.0

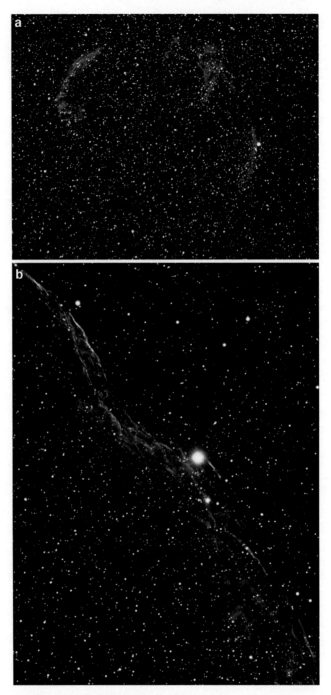

Fig. 9.5. (**a**) Entire Veil Nebula. North is up. (**b**) Western Loop of Veil Nebula: The Witch's Broom. Northwest is up.

September 19: Eastern Loop of Veil: The Network Nebula

Designation	NGC 6992/6995
Other names	Caldwell 33, Network Nebula
Right ascension	20 h 56.4 min
Declination	+31° 43'
Magnitude	–
Size	60 × 8 arcmin
Constellation	Cygnus

Only in the furnace of a supernova explosion can the heavy elements like calcium, gold, and silver be forged. Without these elements, life on earth would never exist. Thus, we owe our very existence to the sacrifice of the supergiant stars that burn so brightly but so briefly.

Imaging. The Veil can be framed with large or small fields of view. With a camera lens or a multiframe mosaic, you can capture the entire Veil with a field of 4 × 3°. The eastern loop (Network Nebula) or the western loop (Witch's Broom) can be framed with moderate size fields of 60 × 40 arcmin. The delicate filaments are revealed best by LRGB methods with a filtered luminance using either an H-alpha filter or a red filter, or both. The western loop is more challenging because of the foreground star of magnitude 4.2 which will bloom severely in a non-antiblooming CCD camera unless you shorten your exposures. Single-shot color cameras will require very long exposures for the Veil.

Processing. The filamentary stands of the Veil Nebula should be the focus of your processing. Consider deconvolution of your filtered luminance to begin sharpening of these filaments. Consider blending your Ha channel lightly (10–20%) into your red channel to create richer red color, but avoid too much Ha that may overpower the green-teal glow of excited oxygen. After color combining, perform routine histogram adjustments with digital development or curves/levels. Restrain overbrightening of an Ha luminance that may overpower your RGB channels. Apply an Ha luminance at less than 100% opacity. An alternative is to blend your Ha channel with your red channel for a smoother luminance. Sharpen the luminance with either smart sharpen, high pass filtering, or unsharp masking (Fig. 9.6).

Fig. 9.6. Eastern Loop of Veil Nebula: The Network Nebula. North is up.

September: Autumn Assortment

Camera	ST10XME
Telescopes	5.5-in. TEC refractor at $f/7$
Field of view	72 × 48 arcmin
Exposures	Luminance clear 6×5 min, Ha 12×5, R 18×5, unbinned
	R 6×5 min, G 5×5 min, B 9×5 min, binned 2×2
Scale	2.0 arcsec/pixel
Limiting magnitude	6.0

September 18: Pelican Nebula

Designation	IC 5070
Other names	Pelican Nebula
Right ascension	20 h 50.8 min
Declination	+44° 21'
Magnitude	–
Size	80 × 60 arcmin
Constellation	Cygnus

The long beak of the Pelican Nebula looks more like the head of a prehistoric Pterodactyl than the head of a bird. The eye is ghostly dark, but the nearby bright star hints of a displaced eyeball. The Pelican Nebula is part of a huge H-II cloud of hydrogen gas, illuminated by nearby stars. Like other H-II regions, the hydrogen gas is excited by the stellar energy, and then emits its own light at the characteristic red wavelength of hydrogen. Located at a distance of 1,500 light-years in the direction of the constellation Cygnus the Swan, the Pelican Nebula is separated from the adjacent larger North American Nebula by a broad band of light-absorbing dark clouds.

Imaging. The Pelican Nebula is framed tightly with a 90 × 60 field of view, oriented north–south in the long axis. An alternative frame is a much larger field of view to include the North American Nebula also (see section "September 20: North American Nebula"). As with other bright nebula, you can use shorter subexposures if needed to compensate for suboptimal tracking and to reduce blooming from the two bright fifth magnitude stars; in this image, I achieved good results with 1-min exposures. The Pelican is not as bright as the North American Nebula yet sufficiently bright for RGB or single-shot color. The depth of red luminosity is better shown with a filtered luminance using either a red filter or an H-alpha filter.

Processing. After image calibration, debloom the stars of magnitude 5 that are located above the beak and on the back of the Pelican. If you have an H-alpha luminance, consider blending it with your red channel using Pixel Math or using other methods before RGB combine to create a stronger red channel and to create a better luminance. As you finish, touch up bloating or color fringing around the brighter stars (Fig. 9.7).

Fig. 9.7. Pelican Nebula. North is up.

Camera	ST10XME
Telescope	4-in. Astro-Physics refractor at *f*/5
Field of view	97 × 65 arcmin
Exposures	Luminance red 83 × 1 min, unbinned R and G each 20 × 1 min, B 27 × 1, binned 2 × 2
Scale	2.8 arcsec/pixel
Limiting magnitude	3.5

September 20: North American Nebula

Designation	NGC 7000
Other names	Caldwell 20
Right ascension	20 h 58.8 min
Declination	+44° 20'
Magnitude	–
Size	120 × 100 arcmin
Constellation	Cygnus

The North American Nebula is easy to recognize by its resemblance to our continent, with Mexico, the Caribbean, and Florida clearly recognized. This geographic shape is of course just coincidence. From a dark sky site, I have observed the distinct shape of this large faint nebula with the aid of special nebula filters that darken the background sky. The nebula lies at a distance of 1,800 light-years in the constellation Cygnus, and spans a region of 50 light-years across.

Imaging. You can choose to frame both the North American Nebula and Pelican Nebula together with a field of view of at least 150 × 150 arcmin. A tighter frame of just the North American Nebula can be achieved with a frame of 120 × 100 arcmin. Although the nebula is bright enough to image with single-shot color or RGB methods, greater detail in the nebula is revealed with a filtered luminance using a red filter or an H-alpha filter. If you are using an H-alpha filter, consider binning your H-alpha channel 2 × 2 if needed to speed up light collection.

Processing. After calibrating your exposures, apply deblooming tools for the brighter stars. Then, align and combine your images. If you are using an H-alpha luminance, blend it with your red channel image to create a new blended channel that can be used for both luminance and red channels. The blending percentages will depend on the length of your exposures, your binning, and the bandwidth of your Ha filter. As a starting point, measure the brightness of an unsaturated star in your red and H-alpha channels. For example, if the star's pixel value measures 5,000 with the H-alpha filter and 20,000 with the red filter, create a blend of 50% red and 200% H-alpha, which will keep the star at 20,000 to preserve star color while boosting red nebulosity (Fig. 9.8).

Fig. 9.8. North American Nebula (*top*) and Pelican Nebula (*bottom*). East is up.

September: Autumn Assortment

Camera	ST10XME
Telescope	Nikon 180-mm camera lens at $f/2.8$
Field of view	235 × 160 arcmin
Exposures	H-alpha 6 × 5 min binned 2 × 2
	R and G each 3 × 5 min, B 4 × 5 min, unbinned
Scale	7.8 arcsec/pixel
limiting magnitude	6.0

September 20: Fetus Nebula

Designation	NGC 7008
Other names	Hidden Treasure 103
Right ascension	21 h 00.6 min
Declination	+54° 33'
Magnitude	10.6
Size	2.5 × 1.5 arcmin
Constellation	Cygnus

The Fetus Nebula, NGC 7008, is a planetary nebula, representing a dying star that has shed its outer layers, similar to the Ring Nebula, Dumbbell Nebula, and Helix Nebula, but more distant, residing 2,800 light-years away. The nebula is comprised of two concentric elliptical shells encompassing 1.3 × 1.0 light-years. For comparison, the Dumbbell Nebula has a similar radius of 1.4 light-years, but is only half the distance of the Fetus Nebula. The magnitude 13 central white dwarf is easy to spot on astrophotographs. In his book "Hidden Treasures," Stephen O'Meara attributes the name "Fetus Nebula" to Eric Honeycutt in a 2001 issue of Amateur Astronomy magazine.

Imaging. Although small, the Fetus Nebula has enough detail to generate visual interest with a small field of view. As with other small targets, wait until a night of steady skies to attempt a high-resolution image. The magnitude 10.6 light is spread out and best captured with luminance layering from a dark sky site. Single-shot color or RGB methods would require longer exposures.

Processing. After image calibration, apply deblooming methods if needed for the bright stars around the nebula. If you have strong luminance exposures, consider applying deconvolution after you have combined your luminance channel. After color combining and adjusting your histogram, apply more sharpening to the nebula. Avoid sharpening the bright foreground stars projecting over the nebula. Enrich color in the nebula by either boosting saturation or using "match color" in Photoshop. If the magnitude 8.5 gold and 9.5 blue double stars at the southern border of the nebula appear bloated, apply star-shrinking methods (see "Final Cleanup" in Chap. 15) (Fig. 9.9).

September: Autumn Assortment

Fig. 9.9. Fetus Nebula. North is up.

Camera	ST10XME
Telescopes	12-in. Meade LX200R at *f*/7
Field of view	22 × 15 arcmin cropped to 16 × 11 arcmin
Exposures	Luminance clear 15 × 5 min, unbinned R 7 × 5 min, G 8 × 5 min, B 6 × 5 min, binned 2 × 2
Scale	0.6 arcsec/pixel
Limiting magnitude	6.0

September 21: Iris Nebula

Designation	NGC 7023
Other names	Caldwell 4
Right ascension	21 h 01.8 min
Declination	+68° 12′
Magnitude	–
Size	10 × 10 arcmin
Constellation	Cepheus

The Iris Nebula floats as a delicate violet flower 1,200 light-years away, illuminated by a young hot star at its core. Brilliant blue starlight reflecting from a cloud of dust provides most of the Iris Nebula's shimmering facade, tempered by faint red luminescence of dust grains energized by intense ultraviolet radiation. Additional dust that is more distant from the central star reflects light only dimly, creating an opaque veil hiding more distant background stars. Astronomers designate this object as NGC 7023, following the visual description used by Johann Dreyer in his "New General Catalogue," published in 1888.

Imaging. The Iris Nebula can be framed either to emphasize just the bright nebula, or with a larger field to emphasize the surrounding dark nebula. The bright areas can be framed nicely with a high-resolution field of view of 20 arcmin. The larger dark nebula can be framed with a field of view of 50 arcmin. The Iris Nebula is dim, and benefits from luminance layering methods. Single-shot color or RGB methods would require longer exposures. As with most dim reflection nebulae, imaging the Iris Nebula is very difficult in areas of light pollution.

Processing. Process the Iris Nebula with routine methods. The bright central star of magnitude 7.3 may bloom severely with a non-antiblooming camera. If so, begin the correction by applying the antiblooming tools after calibration. After image alignment and combining, adjust color balance and then enrich color. The central area of the nebula should appear a rich blue, and the outer regions should fade to a pale violet tone. Sharpen the brighter central areas of the nebula, but avoid sharpening the central star. If the central star is bloating, apply star-shrinking methods (Fig. 9.10).

Fig. 9.10. Iris Nebula. East is up.

September: Autumn Assortment

Camera	ST10XME
Telescope	5.5-in. TEC refractor at f/5
Field of view	72 × 48 arcmin
Exposures	Luminance clear 17 × 5 min, unbinned R 6 × 5 min, G 4 × 5 min, B 9 × 5 min, binned 2 × 2
Scale	2.0 arcsec/pixel
Limiting magnitude	6.0

September 28 and 29: Globular Clusters M15 and M2

Designation	NGC 7078	NGC 7089
Other names	Messier 15	Messier 2
Right ascension	21 h 30.0 min	21 h 33.5 min
Declination	+12° 10′	−00° 49′
Magnitude	6.0	6.4
Size	18 × 18 arcmin	16 × 16 arcmin
Constellation	Pegasus	Aquarius

M2 and M15 are similar globular clusters in the autumn skies. Like other globular clusters, they are essentially islands of hundreds of thousands of suns that were formed at a similar time, about 13 billion years ago at the time of the formation of our galaxy. Globular clusters orbit at the periphery of our galaxy, with almost 150 of them scattered around the Milky Way galaxy. Containing 150,000 stars within a diameter of 175 light-years, M2 is one of the more compact globular clusters. M15 has the same outer diameter, but has a much denser core, possibly harboring a black hole at its center. Both are between 30,000 and 40,000 light-years distant.

Imaging. Globular clusters are excellent targets for beginning astrophotographers. The larger and brighter globular clusters, like M2 and M15, can be framed either with a medium size field of view of 30–40 arcmin or with a smaller field of view for high-resolution imaging. These objects are bright enough to be imaged successfully with either single-shot color, RGB, or LRGB methods. With longer exposures of globular clusters, more of the faint outer stars become revealed, increasing the apparent size of the cluster in the final image.

Processing. See comments for Globular Clusters M3, M5, M12, and M13 (Fig. 9.11).

September: Autumn Assortment

Fig. 9.11. (**a**) Globular Cluster M15. East is up. (**b**) Globular Cluster M2. Northwest is up.

Camera	ST10XME
Telescope	8-in. Celestron Schmidt–Cassegrain at f/10
Field of view	26 × 17 arcmin
Exposures (M15)	Luminance IDAS 20 × 3 min, unbinned R 3 × 3 min, G 3 × 3 min, B 4 × 3 min, unbinned
Exposures (M2)	Luminance IDAS 14 × 5 min, unbinned R and G each 4 × 5 min, B 8 × 5 min, unbinned
Scale	0.7 arcsec/pixel
Limiting magnitude	3.5

September 30: Nebula IC 1396 and the Elephant's Trunk

Designation	IC 1396
Other names	Hidden Treasure 105
Right ascension	21 h 39.1 min
Declination	+57° 30'
Magnitude	3.5
Size	90 × 90 arcmin
Constellation	Cepheus

The Elephant's Trunk is a dark nebula harboring star formation in the western (right) quarter of the large emission nebula IC 1396, at a distance of 2,500 light-years. Magnitude 4 Mu Cepheii, also known as Herschel's Garnet Star, is a red supergiant star at the north (upper) border of IC 1396.

Imaging. IC 1396 is similar to a larger version of the Rosette Nebula, and requires a field of view of about 3° for proper framing. The Elephant's Trunk is the most interesting feature of IC 1396, and is best framed with a smaller field of view of about 30–50 arcmin. The rich features of the nebulosity are enhanced by LRGB with a filtered luminance using a red filter or an H-alpha filter.

Processing. As with other emission nebulae, H-alpha exposures should be blended with the red exposures both for the luminance layer and for the red channel (see sections "September 20: North American Nebula" in the present chapter and "Luminance Layering" in Chap. 15) (Fig. 9.12).

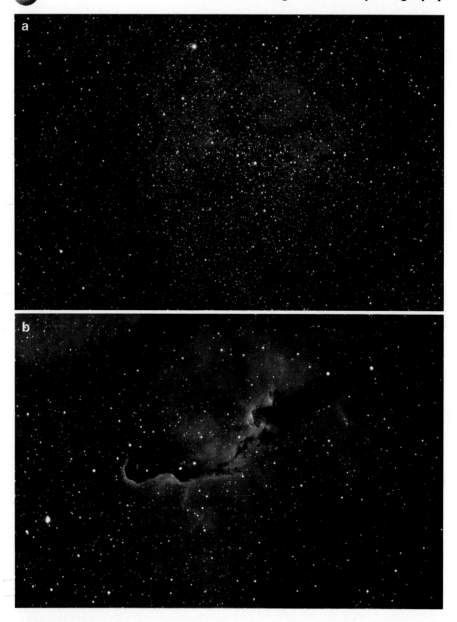

Fig. 9.12. (a) Emission Nebula IC 1396. North is up. (b) Elephant's Trunk. North is up.

September: Autumn Assortment

Camera	ST10XME
Telescope	Nikon 180-mm camera lens at $f/2.8$
Field of view	235 × 160 arcmin
Exposures	H-alpha 6 × 5 min binned 2 × 2
	R, G, and B each 1 × 5 min, unbinned
Scale	7.8 arcsec/pixel
Limiting magnitude	6.0

Camera	ST10XME
Telescope	5.5-in. TEC refractor at $f/5$
Field of view	71 × 48 arcmin
Exposures	Luminance Ha 15 × 10 min, unbinned
	R 10 × 5 min, G and B each 8 × 5 min, binned 2 × 2
Scale	2.0 arcsec/pixel
Limiting magnitude	6.0

CHAPTER TEN

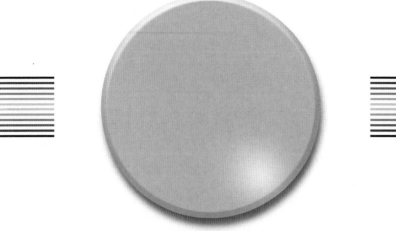

October: Halloween Treats

October 4: Cocoon Nebula

Designation	IC 5146 (cluster), Sh-125 (nebula)
Other names	Caldwell 19
Right ascension	21 h 53.5 min
Declination	47° 16'
Magnitude	7.2 (cluster), 9.3 (nebula)
Size	12 × 12 arcmin
Constellation	Cygnus

The Cocoon Nebula is a stellar nursery, with its delicate fabric of glowing hydrogen torn asunder by the solar wind of newborn stars. Imagine scanning the heavens from a planet within this nebula. While we envision a glowing scarlet sky obscuring most stars, in reality the inhabitants would barely notice the nebula. The faint rustic shade emitted by the hydrogen cloud would just contribute to the background canvas of the night sky. Beyond the central glow of hydrogen, the surrounding interstellar dust begins to dominate. At the edge of the red emission nebula, this dust reflects pale blue light. Farther away, the dust fades, and instead blocks the light of background stars.

Imaging. You have many options for framing the Cocoon Nebula. A small field of view, as in my image, provides the most detail in the nebula. A larger field of view, of about 30 arcmin, shows a greater extent of the surrounding dark nebula. An even larger field of 70 arcmin reveals the dark nebula Barnard 168 extending to the west of the Cocoon Nebula. The Cocoon Nebula is relatively bright, and can be imaged with either LRGB, RGB, or single-shot color methods. An H-alpha luminance enhances the red nebulosity but is not necessary.

Processing. Begin by routinely calibrating, aligning, and combining your images. If your images are high resolution with long exposures, consider deconvolution of a luminance or synthetic luminance channel. Adjust your histogram with either digital development or curves/levels. If applying a red or H-alpha luminance layer, use well under 100% opacity to avoid suppression of the faint blue reflection at the outer border of the red emission nebulosity. Sharpen the brighter areas of the nebula, and smooth the dimmer areas that appear noisy (Fig. 10.1).

October: Halloween Treats

Fig. 10.1. Cocoon Nebula. North is up.

Camera	ST10XME
Telescope	12-in. Meade LX200R at f/7
Field of view	23 × 15 arcmin
Exposures	Luminance Hα 12 × 10 min, binned 2 × 2 R 7 × 5 min, G and B each 6 × 5 min, binned 2 × 2
Scale	1.3 arcsec/pixel
Limiting magnitude	6.0

October 9: Wolf's Cave and the Cepheus Flare

Designations	Van den Bergh 152
Other names	Cederblad 201, Barnard 175
Right ascension	22 h 13.6 min
Declination	+70° 18'
Magnitude	8.8
Size	12 × 6 arcmin
Constellation	Cepheus

Just in time for Halloween, this complex region resembles a ghostly apparition flying upward in this view, with a bright glowing blue heart and dim brown body floating behind. In reality, the bright star at the heart is an interloper to the region, slamming into the dark nebula Barnard 175, causing collisional and ultraviolet excitation, producing the reflection nebula Cederblad 201. Together, the dark and reflection nebulae are called Van den Bergh 152. The term Wolf's Cave is named after Max Wolf who discovered the "long, dark lacuna" in 1908. Since 1934 when Hubble studied this region, it has been called the Cepheus Flare.

Imaging. Van den Bergh 152 can be framed with a small field of view, as in my image, to focus on the bright refection nebula Cederblad 201. A larger field of 50 arcmin shows more of the dark nebula B175, and would be a better choice if your tracking is less than perfect or the seeing is unsteady. This entire region is among the more difficult targets in this book because it is faint, which requires long exposures and dark skies. Luminance layering with a clear filter is suggested to speed up light collection. Do not try this from a light-polluted suburb!

Processing. For this object, use mostly routine luminance layering methods. Because deconvolution can introduce noise, you should skip this step. To grab every photon of data, consider adding your color exposures to your luminance for a stronger luminance. Apply histogram adjustments to reveal the faint brown glow of dust in the dark nebula. Apply smoothing methods to the dark nebula region. Sharpen only the bright region around the reflection nebula Cederblad 201 (Fig. 10.2).

Fig. 10.2. Wolf's Cave and Cepheus Flare. South is up.

October: Halloween Treats

Cameras	ST10XME (luminance, RGB), ST2000XM (RGB)
Telescopes	12-in. Meade LX200R at $f/7$ (luminance, RGB)
	5.5-in. TEC refractor at $f/7$ (RGB)
Field of view	23 × 15 arcmin
Exposures	Luminance clear 43 × 5 min, unbinned
	R 5 × 5 min, G 4 × 5 min, B 6 × 5 min, binned 2 × 2
	R 7 × 10 min, G 6 × 10 min, B 7 × 10 min, unbinned
Scale	0.6 arcsec/pixel
Limiting magnitude	6.0

October 13: Helix Nebula

Designation	NGC 7293
Other names	Caldwell 63
Right ascension	22 h 29.6 min
Declination	–20° 48'
Magnitude	7.3
Size	18 × 18 arcmin
Constellation:	Aquarius

The Helix Nebula, in the constellation Aquarius, surrounds a dying star that has blown off its outer layers, once its central supply of nuclear fuel is nearly exhausted. The remnant central star is a dense "white dwarf" that can no longer support nuclear reactions. At a distance of 450 light-years, it is the closest planetary nebula to earth. The very hot white dwarf ionizes an inner shell of oxygen, which appears green in this image, and an outer shell of hydrogen, which appears red.

Imaging. The Helix Nebula has the largest apparent diameter of any planetary nebula, allowing a moderate size frame of 30–50 arcmin. Orient your frame to place the long axis of the Helix at an angle, unlike the bland parallel position in my image. If you are imaging from northern latitudes, your challenge is from the low altitude of this object. If you are using a clear luminance or a single-shot color camera, obtain these images near the meridian when the Helix is at its maximum altitude, which reduces blurring from atmospheric refraction. For RGB imaging, obtain your blue channels near the meridian to reduce the effects of atmospheric extinction, which affects blue exposures the most. An H-alpha luminance may enrich the peripheral detail at the expense of the teal center.

Processing. Routine processing steps work well for this large bright target. If your combined channels are deep with little noise, adjust your histogram to show the faint outer rim of red nebulosity on the east side of the main nebula. When balancing color, make sure that the central region maintains a proper teal color. Sharpen the inner regions of the red nebulosity to reveal intricate detail in this area. If required by image noise, smooth the central teal region and the faint outer red nebulosity (Fig. 10.3).

October: Halloween Treats

Fig. 10.3. Helix Nebula. Northwest is up.

Camera	ST2000XM
Telescope	5.5-in. TEC refractor at $f/7$
Field of view	41 × 31 arcmin
Exposures	Luminance clear 22 × 10 min, unbinned R 10 × 10 min, G 10 × 10 min, B 5 × 10 min, unbinned
Scale	1.6 arcsec/pixel
Limiting magnitude	6.0

October 14: Stephan's Quintet

Designation	NGC 7317, 7318, 7319, 7320
Other names	Arp 319, Hickson 92
Right ascension	22 h 36.1 min
Declination	+33° 57'
Magnitude	12.0
Size	4 × 3 arcmin
Constellation	Pegasus

Stephan's Quintet, the first compact galaxy group ever described, was discovered by Édouard Stephanis in 1877 at Marseilles Observatory. He identified five tightly grouped galaxies, shown in the center of this image. The lower right galaxy, NGC 7317, is a dull elliptical galaxy. To its left are two converging spiral galaxies, NGC 7318. The fourth, NGC 7319, in the upper left center, is another interacting spiral, whose gravitational interaction with the other galaxies in this group is distorting their spiral arms. The brightest member of the visual group, in the lower center, is the bluish spiral galaxy NGC 7320. At the far left of the image from the main group, a faint barred spiral with a symmetrical ring structure (NGC 7320C) is now considered to be an outlying member of the galaxy cluster. Spectroscopy shows a small redshift for NGC 7320 but a much larger redshift for the other galaxies in the group. Because redshift correlates with distance, NGC 7320 is only 60 million light-years away while the rest of the galaxy cluster is about 300 million light-years distant.

Imaging. If you are trying to get some detail in the individual galaxies, frame Stephan's Quintet with a small field of view under 20 arcmin, using your largest telescope to achieve high resolution under 1 arcsec/pixel. This requires steady skies, accurate tracking and guiding, precise focusing, and long exposures. A more humble approach would be to include Stephan's Quintet as a small element in a larger field of view that includes the Deer Lick Group of galaxies centered only 30 arcmin to the north–northeast; a frame of 30 × 45 arcmin would display both groups nicely. These galaxies are dim and therefore easier to image with luminance layering using a clear luminance.

Processing. After image calibration, apply deconvolution to your luminance channel. Deconvolution works its magic best on high-resolution images such as this that may be limited by atmospheric seeing, but requires an image with rich data and low noise. After combining your channels,

The 100 Best Targets for Astrophotography

Fig. 10.4. Stephan's Quintet Galaxy Cluster. North is up.

October: Halloween Treats

adjust your histogram to reveal the faint spiral arm being pulled off NGC 7319. Sharpen just the bright areas of the galaxies to show their structure. Enhance your color by either increasing saturation or using "match color" in Photoshop to display the blue tint of NGC 7320 compared with the yellowish cores of the other galaxies (Fig. 10.4).

Camera	ST10XME
Telescopes	12 in. Meade LX200R at f/7
Field of view	22 × 15 arcmin cropped to 13 × 10 arcmin
Exposures	Luminance clear 29 × 5 min, unbinned R 10 × 5 min, G 10 × 5 min, B 14 × 5 min, binned 2 × 2
Scale	0.6 arcsec/pixel
Limiting magnitude	6.0

October 15: Deer Lick Galaxy Group

Designations	NGC 7331	NGC 7335	NGC 7337
Other names	Caldwell 30		
Right ascension	22 h 37.1 min	22 h 37.3 min	22 h 37.4 min
Declination	+34° 25'	+34° 27'	+34° 22'
Magnitude	9.5	14.5	15.7
Size	11 × 4 arcmin	1.7 × 0.6 arcmin	1.3 × 0.8 arcmin
Constellation	Pegasus	Pegasus	Pegasus

The Galaxy NGC 7331, at a distance of 46 million light years, is the brightest member of a galaxy group called the Deer Lick group. Smaller members of this group are on the left side of this image, including the smaller spiral galaxies NGC 7335 (upper left) and NGC 7337 (lower left), which are probably ten times farther away than NGC 7331. NGC 7331 was discovered in 1784 by William Herschel, and was one of the brightest galaxies overlooked by Messier in his catalog. It appears nearly edge on, tilted at an inclination of 77 degrees. Its structure is remarkably similar to our own Milky Way Galaxy, with a comparable overall mass, spiral structure, distribution of stars, and central supermassive black hole.

Imaging: If you want to focus just on the largest galaxy in the Deer Lick Group, then you can frame NGC 7331 with a field of view as small as 15 x 10 arcminutes. If you want to include several surrounding small galaxies then you should choose a larger frame of up to 25 arcminutes. A third choice would be a large field of 30 x 45 arcminutes to include both the Deer Lick Group of galaxies and Stephan's Quintet, which is centered only 30 arcminutes to the south-southwest. NGC 7331 is sufficiently bright for either RGB methods or single shot color, but luminance layering with a clear filter will help reveal finer structure in its spiral arms and more detail in the smaller galaxies.

Processing. Try to optimize detail in the spiral arms of NGC 7331 and the smaller galaxies. After calibrating, aligning, and combining your images into channels, consider the deconvolution of your luminance. After color combine and histogram adjustment, apply more sharpening to the brighter galactic structures (Fig. 10.5).

Fig. 10.5. Deer Lick Galaxy Group. North is up.

Camera	ST10XME
Telescopes	12-in. Meade LX200R at $f/7$
Field of view	22 × 14 arcmin
Exposures	Luminance clear 27 × 5 min, unbinned
	R, G, and B each 6 × 5 min, binned 2 × 2
Scale	0.6 arcsec/pixel
Limiting magnitude	6.0

October 19: Flying Horse Nebula

Designation	NGC 7380, Sh2-142
Other names	Hidden Treasure 106
Right ascension	22 h 47.9 min
Declination	+58° 06'
Magnitude	7.2
Size	20 × 20 arcmin
Constellation	Cepheus

NGC 7380 is a young cluster, between 2 and 4 million years old, containing about 125 bright type O and B stars near the center of this image. The red emission nebula Sh2-142 is part of the larger H-II region illuminated by these stars. Interspersed dust lanes shape the contours of the nebula into a flying horse facing to the right, trailing a long flowing cape extending to the upper left. At the base of the horse's tail, an eclipsing binary star DH Cephei is the source of much of the ionization in the nebula. This complex resides at a distance of 9,700 light-years, at the edge of the Cepheus OB1 association.

Imaging. The Flying Horse Nebula can be framed by a field of view between 25 and 50 arcmin. The extent of the nebulosity is best shown using a filtered luminance with either an H-alpha filter or a red filter. Single-shot color cameras and RGB methods can be successful with longer exposures, but may be dominated by many bright stars in NGC 7380.

Processing. Begin with routine image calibration and alignment, followed by combining your color channels. If you have obtained an H-alpha channel, consider blending it with your red channel to create a new stronger red channel that will better match your H-alpha luminosity. After routine color combine, adjust your histogram with either digital development or curves/levels. Gently sharpen the brighter areas of nebulosity, and smooth the dimmer areas of nebulosity that may show noise. If you are applying an H-alpha luminance, either first soften the H-alpha channel by blending it with a copy of the red channel, or apply the luminance layer at less than 100% opacity. Enrich the red nebula and blue bright stars by either increasing saturation or using Photoshop's "match color" (Fig. 10.6).

Fig. 10.6. Flying Horse Nebula NGC 7380. South is up.

Camera	ST2000XM
Telescopes	5.5-in. TEC refractor at $f/7$
Field of view	41 × 31 arcmin
Exposures	Luminance Ha 16 × 20 min, unbinned R 13 × 10 min, G 8 × 10 min, B 10 × 10 min, binned 2 × 2
Scale	1.6 arcsec/pixel
Limiting magnitude	6.0

October 20: Cave Nebula

Designation	Caldwell 9
Other names	Sh2-155
Right ascension	22 h 56.8 min
Declination	+62° 37'
Magnitude	10.0
Size	50 × 30 arcmin
Constellation	Cygnus

The Cave Nebula is a large emission nebula of glowing hydrogen gas at a distance of 2,800 light-years, which is part of a larger Cepheus B giant molecular cloud. The young stars which are emerging from this cloud are also energizing and illuminating the nebula. The dark "cave" is caused by interstellar dust, obscuring the light from both the nebula and background stars.

Imaging. The Cave Nebula is framed well with a variety of fields of view ranging from 30 to 60 arcmin. The "cave" itself measures about 15 arcmin, and therefore can be framed with more or less surrounding nebulosity. The nebulosity is faint and spread out, and therefore is best revealed by a layered luminance using either a red or H-alpha luminance. RGB methods and single-shot color would require longer exposures.

Processing. Luminance layering methods work best for the Cave Nebula. After aligning and combining your exposures, consider blending the H-alpha with the red channel to improve the intensity of red nebulosity, and then perform color combine. Adjust your histograms with either digital development or curves/levels to reveal the nebulosity. If you are using an H-alpha luminance, then apply the luminance layer at less than 100% opacity. Sharpen the brighter areas of nebulosity. Apply smoothing to the dim areas of nebulosity that may show more noise. If needed, enhance red nebulosity with either a boost in color saturation or using "match color" in Photoshop (Fig. 10.7).

Fig. 10.7. Cave Nebula. East–northeast is up.

Camera	ST10XME
Telescope	5.5-in. TEC refractor at f/5
Field of view	72 × 48 arcmin
Exposures	Luminance Ha 13 × 5 min, unbinned R, G, and B each 4 × 5, binned 2 × 2
Scale	2.0 arcsec/pixel
Limiting magnitude	6.0

October 26: Bubble Nebula and M52 Cluster

Designation	NGC 7625	NGC 7654
Other names	Caldwell 11	Messier 52
Right ascension	23 h 20.7 min	23 h 24.2 min
Declination	+61° 12'	+61° 36'
Magnitude	10.0	6.9
Size	15 × 8 arcmin	16 × 16 arcmin
Constellation	Cassiopeia	Cassiopeia

The M52 cluster contains hundreds of young stars that formed together about 20 million years ago. The Bubble Nebula is shaped by the powerful solar winds from a Wolf–Rayet star impacting a giant molecular cloud. The two objects are at similar distances from Earth: M52 at 5,100 light-years and the Bubble Nebula at 7,100 light-years.

Imaging. You can frame both the Bubble Nebula and the cluster M52 using a large field of view. If your sky is steady and your tracking is accurate, you can use a smaller field of view for high-resolution imaging of the Bubble Nebula (Fig. 10.8).

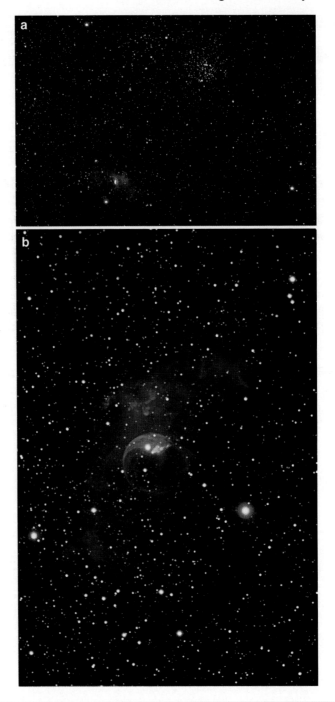

Fig. 10.8. (a) Bubble Nebula and Open Cluster M52. East is up. (b) Bubble Nebula. North is up.

October: Halloween Treats

Camera	ST10XME
Telescope	4-in. Astro-physics refractor at *f*/6
Field of view	79 × 55 arcmin
Exposures	Luminance IDAS 25 × 3 min, unbinned R 5 × 3 min, G 5 × 3 min, B 6 × 3 min, binned 2 × 2
Scale	2.3 arcsec/pixel
Limiting magnitude	3.5

Camera	ST10XME
Telescope	12-in. Meade LX200R at *f*/7
Field of view	23 × 15 arcmin
Exposures	R 15 × 5, G 11 × 5, B each 12 × 5 min, binned 2 × 2 Red channel also used for luminance at 75% opacity
Scale	1.3 arcsec/pixel
Limiting magnitude	6.0

October 27: Blue Snowball Nebula

Designation	NGC 7662
Other names	Caldwell 22
Right ascension	23 h 25.9 min
Declination	+42° 33'
Magnitude	8.3
Size	0.6 × 0.6 arcmin
Constellation	Andromeda

The Blue Snowball is a planetary nebula, representing a dying star that has shed its outer layers in a series of shells. The hot glowing ember of the collapsed star survives as a white dwarf at the center of the nebula, energizing the concentric shells with fierce ultraviolet radiation. Through a large telescope, the Blue Snowball appears visually as an oval turquoise gem against a velvet black sky. This teal color arises from the emission lines of doubly ionized Oxygen (O-III) at wavelengths of 496 and 501 nm. At a distance of 3,200 light-years, it is slightly closer to earth than the Little Dumbbell M76.

Imaging. The Blue Snowball Nebula is the smallest target in this book, with a diameter under 1 arcmin. To extract any detail, high resolution is required. Use a large telescope with a sufficient focal length to image at under 1 arcsec/pixel (see Table 14.1). Wait for a night of exceptional seeing. If your tracking is less than perfect, consider shorter subexposures to reduce tracking errors. Luminance exposures may be helpful to reduce exposure times further if your tracking is introducing any blur into the image. The magnitude 8.3 light of the Blue Snowball is concentrated in a small area, which allows single-shot color or RGB methods to be used.

Processing. Begin processing by calibrating, aligning, and combining your exposures into color channels. If you are using RGB methods, consider creating a synthetic luminance by combining all of your exposures into a new channel. Apply deconvolution to your (synthetic) luminance channel. Adjust your histogram with curves/levels; avoid digital development for this object. Apply aggressive sharpening to the nebula in your luminance layer. Enrich color in your color layer by either boosting saturation or using "match color" in Photoshop (Fig. 10.9).

Fig. 10.9. Blue Snowball. North is up.

Camera	ST10XME
Telescopes	12-in. Meade LX200R at f/7
Field of view	22 × 15 arcmin cropped to 12 × 8 arcmin
Exposures	R 10 × 3 min, G 9 × 3 min, B 8 × 3 min, unbinned
Scale	0.6 arcsec/pixel
Limiting magnitude	6.0

CHAPTER ELEVEN

November: The Great Galaxies

November 1: Andromeda Galaxy and Companions

Designation	NGC 224	NGC 221	NGC 205
Other names	Messier 31	Messier 32	Messier 110
Right ascension	00 h 42.7 min	00 h 42.7 min	00 h 40.4 min
Declination	+41° 16'	+40° 52'	+41° 41'
Magnitude	4.4	8.1	7.9
Size	178×63 arcmin	9×7 arcmin	20×12 arcmin
Constellation	Andromeda	Andromeda	Andromeda

The Great Andromeda Galaxy, M31, is the farthest object that most of us will ever see with the naked eye. At a distance of over 2 million light-years, this galaxy is so huge that it occupies an area in the sky several times larger than the full moon! Although similar to our own Milky Way Galaxy, M31 may be twice as large, containing 300 billion suns. Billions of stars are packed tightly together at the galaxy's core, creating the bright central glow, concealing a central massive black hole. Two satellite galaxies of M31 are in the same view. M32, a dwarf elliptical galaxy containing a mere 3 billion solar masses, is tucked in tightly on the lower left of M31. M110 is slightly larger, seen in the upper right of this image.

Imaging. The Andromeda Galaxy should be one of the first targets attempted by a novice imager. If your CCD chip is small, acquire your image through a camera lens to capture a wide field of about 2°. Routine RGB imaging or single-shot color imaging at maximum resolution is easy with this bright large object. Consider LRGB only to reveal the dimmer portions of the outer arms with luminance exposures. Advanced imagers may try mosaics through longer focal lengths to boost the resolution.

Processing. The interesting dark lanes of the Andromeda Galaxy are largely in the dimmer areas. Stretch your histogram with either digital development or levels and curves. Sharpen around the dark lanes to emphasize their contrast. If you have saturated the core, avoid sharpening in the center of the core, or you may end up with abrupt transitions in brightness where you would prefer a smoother appearance. Enrich your color gently to distinguish the yellow core from the brown dark lanes and the blue outer arms. Subtract light-pollution gradients if needed (Fig. 11.1).

November: The Great Galaxies

Fig. 11.1. Andromeda Galaxy M31. North is up.

Camera	ST10XME
Telescope	3.5-in. Takahashi refractor at f/4.5
Field of view	126 × 85 arcmin
Exposures	Luminance clear filter 15 × 5 min, unbinned R 14 × 5 min, G and B each 8 × 5 min, unbinned
Scale	3.5 arcsec/pixel
Limiting magnitude	6.0

November 2: Skull Nebula and Galaxy NGC 255

Designations	NGC 246	NGC 255
Other names	Caldwell 56	
Right ascension	00 h 47.0 min	00 h 47.7 min
Declination	–11° 53'	–11° 28'
Magnitude	10.9	11.7
Size	5 × 4 arcmin	3 × 2.5 arcmin
Constellation	Cetus	Cetus

NGC 246 is a dim planetary nebula in the constellation Cetus the whale, at a distance of 1,800 light-years. Although it is much larger than other better-known planetary nebulae such as the Ring Nebula, it is low in the sky for northern observers, and relatively dim, so it gets much less attention. Like other planetary nebulae, NGC 246 is a dying star that has shed its outer layers, of teal oxygen and red hydrogen, illuminated by the glowing white-hot ember of its fading central star. Through the camera, NGC 246 contrasts with the far more distant barred spiral galaxy NGC 255 on the right, located a whopping 70 million light-years away.

Imaging. The Skull Nebula can be imaged by itself at high resolution using a long focal length, or it can be framed along with NGC 255. If atmospheric seeing is unsteady, or if your tracking is less than ideal, the wider field view will be more rewarding. An LRGB approach helps to image the dim nebula, but at northern latitudes, atmospheric refraction may cause some blurring of a clear luminance.

Processing. Even large professional telescopes show limited detail in the nebula, so do not be disappointed if you cannot sharpen details well. Adjust your color balance if atmospheric extinction suppresses your blue channel (Fig. 11.2).

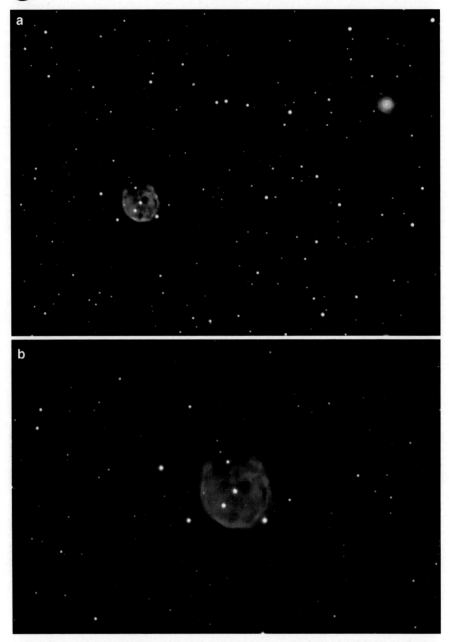

Fig. 11.2. (a) Planetary Nebula NGC 246 with galaxy NGC 255. East is up. (b) Planetary Nebula NGC 246 at high resolution. East is up.

November: The Great Galaxies

Camera	ST10XME
Telescope	5.5-in. TEC refractor at f/7 (both objects)
Field of view	52 × 35 cropped to 43 × 32 arcmin
Exposures	Luminance IDAS 17 × 5 min, unbinned R and G each 4 × 5 min, B 6 × 5 min, binned 2 × 2
Scale	1.4 arcsec/pixel
Limiting magnitude	6.0

Camera	ST10XME
Telescope	12-in. Meade LX200R at f/7 (just NGC 246)
Field of view	23 × 15 arcmin
Exposures	Luminance clear 15 × 5 min, unbinned R 9 × 5 min, G 8 × 5 min, B 12 × 5 min, binned 2 × 2
Scale	0.6 arcsec/pixel
Limiting magnitude	6.0

November 17: Sculptor Galaxy

Designation	NGC 253
Other names	Caldwell 65, Silver Coin
Right ascension	00 h 47.6 min
Declination	–25° 17'
Magnitude	7.1
Size	25 × 7 arcmin
Constellation	Sculptor

Galaxy NGC 253 surely would have been included in Messier's Catalog had it not been so low in the southern sky. At a distance of about 10 million light-years, NGC 253 is the brightest galaxy outside of our Local Group of Galaxies, belonging to the Sculptor Group of Galaxies. This spiral galaxy is tipped steeply to our line of sight, concealing its spiral arms. Caroline Herschel, sister of the more famous William Herschel, discovered this object in 1783.

Imaging. From southern skies, NGC 253 is a spectacular object for imaging. From far northern latitudes, it is a challenge. In fact, this is the most southerly object in the Best Targets. If you are at a northern latitude, atmospheric refraction will bend blue light several arcseconds more than red, so a clear luminance image may show some loss of sharpness and slightly elongated stars. Traditional RGB imaging works best. Also, atmospheric extinction will diminish blue light, becoming worse near the horizon. Therefore, acquire your blue channels near the meridian and your red at lower altitudes. Although bright enough for single-shot color, atmospheric refraction may create color fringes at the edges of stars.

Processing. Even from a dark sky site, the low position of NGC 253 is more likely to introduce significant gradients in the image. After routine aligning, combining, adjusting curves, balancing color, and boosting color, work on correcting the gradients (see section "Dealing Out Gradients" in Chap. 15). For this image, I used GradientXTerminator plug-in for Photoshop to clean up the gradients, using the standard techniques in the program's tutorial. Once your gradients are gone, try aggressive sharpening in the bright areas of the galaxy and noise reduction in the background. If the bright field stars are distracting, gently shrink the offending stars with either a minimum filter with a setting of "1" or other subtle methods (see section "Final Cleanup" in Chap. 15) (Fig. 11.3).

November: The Great Galaxies

Fig. 11.3. Sculptor Galaxy NGC 253. North is up.

Camera	ST2000XM
Telescope	5.5-in. TEC refractor at f/7
Field of view	39 × 28 arcmin
Exposures	B and G each 8 × 10 min, R 13 × 10 min, unbinned
Scale	1.5 arcsec/pixel
Limiting magnitude	6.0

November 18: Pacman Nebula

Designation	NGC 281
Other names	IC 11, Hidden Treasure 3
Right ascension	00 h 52.9 min
Declination	+56° 37′
Magnitude	7.0
Size	35 × 30 arcmin
Constellation	Cassiopeia

The Pacman Nebula lies at a distance of about 10,000 light-years. A bright group of stars in the center illuminate the region of hydrogen gas, glowing red. Dark clouds of dusk obscure some of the glowing gas, creating the "mouth" and "eye" of the Pacman. Star formation occurs in these dense clouds of dust. Because the nebula is in the direction of the Milky Way, innumerable stars glitter in the background and foreground of the image.

Imaging. Try to frame the Pacman Nebula with a field of view between 40 and 60 arcmin. As a moderately large and bright emission nebula, NGC 281 is relatively easy to image with either a single-shot color camera or a monochrome CCD with RGB filters. A clear-filtered luminance is not recommended. If you image from a light-polluted suburb, obtaining your luminance data with an H-alpha filter can help you capture rich nebulosity while suppressing light pollution. From a dark sky site, a red luminance can yield a better match to the stars in your RGB image. If you image unbinned, your red channel can double for both the luminance and red channel, in which case you will want to obtain more red exposures. The high declination of the object allows imagers at northern latitudes to obtain long, deep exposures.

Processing. If you have an H-alpha luminance, consider blending the H-alpha channel with your red channel to better match your RGB data (see section "Luminance Layering" in Chap. 15). After stretching your luminance to reveal the extent and depth of the nebula, use sharpening methods to enhance the contrast and detail, especially around the edges of the dark nebula. If you have a routine RGB image, and you use Photoshop for sharpening, you can immediately fade the sharpening to affect just the luminance component of your image (edit, fade filter, mode: luminance) (Fig. 11.4).

Fig. 11.4. Pacman Nebula NGC 281. North is up.

November: The Great Galaxies

Camera	ST10XME
Telescope	5.5-in. TEC refractor at f/5
Field of view	71 × 47 arcmin
Exposures	Luminance R 24 × 5 min, Hα 9 × 10 min, unbinned R 9 × 5 min, G 10 × 5 min, B 15 × 5 min, binned 2 × 2
Scale	2.0 arcsec/pixel
Limiting magnitude	6.0

November 25: ET Cluster

Designation	NGC 457
Other names	Caldwell 13, Owl Cluster
Right ascension	01 h 19.1 m
Declination	+58° 20'
Magnitude	6.4
Size	20 × 20 arcmin
Constellation	Cassiopeia

NGC 457 had been termed the Owl Cluster, but 20 years ago earned the name the E.T. cluster. Can you see the two large eyes above the thin body with widely stretched arms? (Fig. 11.5). The eyes are formed by double star Phi Cassiopeiae, whose fifth and seventh magnitude components are separated by 134 arcsec. The eyes are probably foreground stars at a distance of 2,000 light-years, with the rest of the cluster closer to 10,000 light-years. Either way, it is a long distance call for E.T to phone home!

Imaging. Frame NGC 457 with at least a 30-arcmin field of view to help distinguish the cluster stars from the rich background stars of the Milky Way. This bright cluster can be imaged nicely with either routine RGB technique or with a single-shot color camera. Phi Cassiopeiae will quickly saturate and bloom in a non-antiblooming CCD, so that shorter exposures may be needed. If you are using a single-shot color camera, wait to image until the ET cluster is above 30° elevation to prevent atmospheric dispersion from blurring your images. If you are using RGB filters through a semi-apochromatic refractor, be sure to refocus between filter changes, so that your star images remain sharp.

Processing. After image calibration, use deblooming tools if needed for Phi Cassiopeiae, prior to image alignment. Alignment is the critical stage in the processing of open clusters. Imperfect alignment creates asymmetrical color halos which ruin the impact of the final image. Apply curves and levels gently to reveal more stars. Do not be afraid to let the brightest stars bloat. Although large stars are distracting in images of nebulae and galaxies, in open clusters a range of star sizes adds visual depth that correlates with differences in star brightness. After correcting the color balance, complete the processing with a boost in color saturation to emphasize the difference between bright blue-white stars and scattered red giants.

November: The Great Galaxies

Fig. 11.5. ET Cluster NGC 457. South is up.

Camera	ST2000XM
Telescope	5.5-in. TEC refractor at f/7
Field of view	41 × 31 arcmin
Exposures	R, G, and B each 18 × 5 min, unbinned
Scale	1.6-arcsec/pixel
Limiting magnitude	6.0

November 29: Triangulum Galaxy

Designation	NGC 598
Other names	Messier 33
Right ascension	01 h 33.8 min
Declination	+30° 39′
Magnitude	6.3
Size	62 × 39 arcmin
Constellation	Triangulum

The Triangulum Galaxy M33 floats 2.3 million light-years away, at the same distance as the Great Andromeda Galaxy. Its diameter of 50,000 light-years is only a third of the Andromeda Galaxy, so it appears substantially smaller through binoculars and telescopes. Nonetheless, even a small telescope at a dark site can detect the spiral arms. In this detailed photograph, one can see an emission nebula, called NGC 604, at the 12:00 position from the nucleus of the galaxy. This region of hydrogen gas glows red, illuminated by 200 young huge bright stars at its center. Although the nebula seems small in this image, NGC 604 is intrinsically 50 times larger than the famous Orion Nebula, but resides 2,000 times farther away within the Triangulum galaxy.

Imaging. My image of the Triangulum Galaxy uses the minimum field of view, of about 50 × 35 arcmin, for a suitable frame. This galaxy is a pleasure to image: large, bright, and rich in both texture and color. Successful images can be obtained with RGB technique, single-shot CCD imaging, or digital cameras. The richest details are possible with LRGB techniques (see section "Luminance Layering" in Chap. 15). The Triangulum Galaxy has many bright H-II regions that can be further enhanced with additional luminance data with either a red filter or an H-alpha filter.

Processing. Routine LRGB methods work well. If you have a component of red or H-alpha luminance, blend it with the clear luminance to enrich the H-II regions. If you have a routine RGB image, you can experiment by applying your red channel as either a luminance or a lighten layer, at well under 50% opacity, and then applying curves to this red layer to enrich the nebulae. Sharpen aggressively in the core, and less in the spiral arms. Enhance color to your taste (Fig. 11.6).

Fig. 11.6. Triangulum Galaxy M33. Northeast is up.

November: The Great Galaxies

Camera	ST10XME
Telescope	5.5-in. TEC refractor at f/7
Field of view	52 × 35 arcmin
Exposures	Luminance clear 18 × 5 min, R 10 × 5 min unbinned R and G each 8 × 5 min, B 7 × 5 min, binned 2 × 2
Scale	1.4 arcsec/pixel
Limiting magnitude	6.0

November 30: Spiral Galaxy M74

Designation	NGC 628
Other names	Messier 74
Right ascension	01 h 36.7 min
Declination	+15° 47'
Magnitude	9.4
Size	10 × 10 arcmin
Constellation	Pisces

Galaxy M74 is one of the great spiral galaxies seen face-on, at a distance of 30 million light-years. Its 100 million stars encompass some bright blue clusters and red hydrogen clouds in the outer arms, compared to older and more tightly packed yellow stars in its core. Dust lanes outline the spiral arms, similar to our own Milky Way Galaxy. The spiral pattern is not caused by the revolution of stars around the galactic disk. Rather, current theory suggests that spiral structure is due to spiral density waves. These density waves compress interstellar gas to stimulate star formation in a spiral pattern.

Imaging. Frame M74 with a small field of view in the range of 20 arcmin. The galaxy is bright enough to image in one evening with either single-shot color, routine RGB, or LRGB methods. The image on the facing page was acquired with less than 2 h of exposures. However, this galaxy has dim spiral arms that will reveal more detail with longer exposures. The best images of M74 are obtained with luminance layering under dark skies. Try to devote several hours, perhaps over several nights, to get rich imaging data. If you are using a monochrome CCD with filters, use nights of substandard seeing to collect the lower resolution binned exposures for your color channels. Collect your higher resolution luminance channel on nights with steady skies.

Processing. Begin processing routinely with calibration, alignment, and combination of exposures into channels. If you have long, deep luminance exposures, try deconvolution. After combining your color channels, enrich your color either by boosting saturation or using match color in Photoshop. Sharpen gently around the core and inner spiral arms, but avoid oversharpening in the dim outer arms (Fig. 11.7).

November: The Great Galaxies

Fig. 11.7. Spiral Galaxy M74. North is up.

Camera	ST10XME
Telescope	12-in. Meade LX200R at f/7
Field of view	23 × 15 arcmin
Exposures	Luminance clear 12 × 5 min, unbinned R 3 × 5 min, G 3 × 5 min, B 3 × 5 min, binned 2 × 2
Scale	0.6 arcsec/pixel
Limiting magnitude	6.0

CHAPTER TWELVE

December: Celestial Potpourri

December 1: Little Dumbbell Nebula

Designation	NGC 650
Other names	Messier 76
Right ascension	01 h 59.3 min
Declination	+51° 34'
Magnitude	10.1
Size	4 × 3 arcmin
Constellation	Perseus

The Little Dumbbell Nebula M76, like other planetary nebulae, shows a unique pattern of red hydrogen and blue-green oxygen gaseous shells blown away from the central star. It is just as large as its namesake the Dumbbell Nebula, but resides about five times farther away at a distance of 4,000 light-years. The Little Dumbbell shows a central bar of gas, surrounded by symmetrical semicircular lobes of faint gas. Many planetary nebulae show bilateral symmetry, which in some cases may be related to a binary system.

Imaging. Many planetary nebulae are small objects yet rich in color, justifying their inclusion among the Best Targets. A long focal length telescope is suggested for the Little Dumbbell Nebula to yield a small imaging scale and thus high resolution. Steady seeing, accurate tracking, and sharp focusing help to display the clear detail. Routine RGB imaging and single-shot color cameras can yield good results, because most of the faint light of the Little Dumbbell is concentrated into a small area. Luminance layering with a clear filter can help to reveal the dimmer components of the nebula.

Processing. Small planetary nebulae need as much detail as possible, because they have to hold the viewer's interest despite appearing almost lost in the center of an image. Try deconvolution methods after combining your images, but before digital development (DDP) or applying curves. Sharpen the nebula again after your histogram adjustments have revealed the outer lobes of the Little Dumbbell. The central bar is brighter, and thus can tolerate more sharpening than the faint outer loops of gas. Adjust your color balance to show the proper teal color of oxygen and deep red color of hydrogen (Fig. 12.1).

December: Celestial Potpourri

Fig. 12.1. Little Dumbbell Nebula M76. North is up.

Camera	ST10XME
Telescope	12-in. Meade LX200R at $f/7$
Field of view	23 × 15 arcmin cropped to 15 × 11 arcmin
Exposures	Luminance clear 17 × 5 min, unbinned R, G, and B each 6 × 5 min, binned 2 × 2
Scale	0.6 arcsec/pixel
Limiting magnitude	6.0

December 5: Nautilus Galaxy NGC 772

Designations	NGC 772
Other names	Hidden Treasure 8
Right ascension	01 h 59.3 min
Declination	+19° 00'
Magnitude	9.9
Size	7 × 5 arcmin
Constellation	Aries

NGC 772 is both huge and distant. Despite its distance of 106 million light-years in the constellation Aries, NGC 772 is twice the diameter of our Milky Way Galaxy, allowing us to see the detailed structure. The asymmetric spiral arms are distorted by interaction with the dwarf elliptical galaxy NGC 770 at the 4:00 position. Stars stewn outward from NGC 770 appear as a fine haze extending away from the pair. I find the shape of NGC 772 resembling the nautilus shell, so I have called it the Nautilus Galaxy. Many background galaxies are visible in this image.

Imaging. NGC 772 is one of the smallest galaxies in the Best Targets. Its unique shape justifies the effort required for imaging. Although single-shot color cameras and routine RGB methods can be successful, LRGB techniques are ideal for this object. Wait until a night of excellent seeing when you have several hours to devote to the luminance alone. The color channels can be obtained on another night when the sky is less steady. Include the smaller elliptical galaxies to the south and southwest in your frame.

Processing. Routine luminance layering methods work well with this object. If you have long, deep luminance exposures, try deconvolution on your luminance channel to begin sharpening the galaxy. During histogram adjustments with either DDP or curves and levels, try to reveal the faint detached outer spiral arm seen on the left side of my image. The central region of NGC 772 is bright and detailed, and can tolerate some aggressive sharpening if your data has a good signal-to-noise ratio. Avoid introducing noise in the region between NGC 772 and NGC 770. If you have light-pollution gradients, be sure not to suppress the smaller elliptical galaxies as you clean up the background (Fig. 12.2).

Fig. 12.2. Nautilus galaxy NGC 772. Northwest is up.

December: Celestial Potpourri

Camera	ST10XME
Telescope	12-in. Meade LX200R at $f/7$
Field of view	22 × 15 arcmin
Exposures	Luminance clear 32 × 5 min, unbinned R and G each 10 × 5 min, B 15 × 5 min, binned 2 × 2
Scale	0.6 arcsec/pixel
Limiting magnitude	6.0

December 10: Double Cluster

Designation	NGC 869/884
Other names	Caldwell 14
Right ascension	02 h 20.0 min
Declination	+57° 08'
Magnitude	4.3
Size	60 × 30 arcmin
Constellation	Perseus

Balance and harmony in the celestial symphony? No objects fulfill this promise better than the Double Cluster in Perseus. About 13 million years ago, 300 young suns emerged in each of the two groups from a single large cloud of gas and dust. Without a doubt, these are among finest objects accessible to small telescopes. The two clusters glitter as a pair of matched celestial jewels, even more beautiful through a telescope than any image can capture. Despite their distance of 7,300 light-years from Earth, they are clearly visible to the naked eye from suburban or rural areas, glowing as a faint patch of light midway between the constellations Cassiopeia and Perseus. In this image, young bright type O and B stars dominate, shining blue-white. A few stars have already burned through their hydrogen, bloating into red giants.

Imaging. The Double Cluster is an excellent target for the beginner, and can be imaged well with either single-shot color cameras or RGB technique. It is bright and large, reducing demands on tracking. Use a short focal length to create a frame at least 60 × 90 arcmin, which allows the clusters to be distinguished from the rich background stars.

Processing. As with all star clusters, precise alignment of the color channels is critical to keep star colors centered without asymmetric halos. When you apply curves and levels, allow the brighter stars to bloat slightly to enhance the apparent differences in magnitude. Gently sharpen the stars. If sharpening creates unnatural color rings around your stars, apply a Gaussian blur, and then fade the blur to the color channel (in Photoshop: edit, fade Gaussian blur, mode:color). Once the rings are suppressed, you can boost star colors by increasing saturation or using "match color" (Fig. 12.3).

December: Celestial Potpourri

Camera	ST2000XM
Telescope	3.5-in. Takahashi refractor at $f/4.5$
Field of view	100 × 75 arcmin
Exposures	R, G, and B each 15 × 2 min, unbinned
Scale	3.8 arcsec/pixel
Limiting magnitude	6.0

Fig. 12.3. Double Cluster in Perseus. East is up.

December 11: Outer Limits Galaxy

Designation	NGC 891
Other names	Caldwell 23
Right ascension	02 h 22.6 min
Declination	+42° 21'
Magnitude	9.9
Size	14 × 3 arcmin
Constellation	Andromeda

Tilted only 6° from edge-on, the spiral galaxy NGC 891 appears as a relatively thin disk. The dark lane bisecting the galaxy is a huge cloud of dust encircling the disk of the galaxy, which blocks out the central band of stars. Its diameter of 110,000 light-years is similar to our own Milky Way, with estimates of its distance varying from 24 to 32 million light-years away. For those of you who remember the TV show "Outer Limits," you may recall this galaxy shown in the opening credits. For this reason, NGC 891 is often called the Outer Limits Galaxy.

Imaging. Frame NGC 891 with a small field of view of about 20 arcmin. This galaxy is deceptively difficult to image well. Extracting detail requires steady seeing, accurate tracking, sharp focus, and long exposures. However, the dark lane is easy to display even under suboptimal conditions, so do not hesitate to have fun trying to image this object. You will want to return as your skills and conditions improve. Use LRGB technique, and devote about half of your time to the color channels to reveal the yellow and brown tendrils of dust along the dark lane, and the faint blue tips of the spiral arms.

Processing. Routine luminance layering techniques work well for this object. If you have rich luminance data, begin sharpening with deconvolution of your luminance channel. After histogram adjustments with either DDP or curves and levels, aggressively sharpen along the dark lane. Gentle sharpening can also be applied to the color channels in the brighter regions around the central dark lane, to emphasize the color in the dusty tendrils. Avoid sharpening along the dimmer borders of the galaxy which can introduce excessive noise (Fig. 12.4).

December: Celestial Potpourri

Fig. 12.4. Outer Limits Galaxy NGC 891. North is up.

Camera	ST10XME
Telescope	12-in. Meade LX200R at $f/7$
Field of view	23 × 15 arcmin
Exposures	Luminance clear 24 × 5 min, unbinned R and B each 12 × 5 min, G 15 × 5 min, binned 2 × 2
Scale	0.6 arcsec/pixel
Limiting magnitude	6.0

December: Celestial Potpourri

December 11: Barred Spiral Galaxy NGC 925

Designation	NGC 925
Other names	
Right ascension	02 h 27.3 min
Declination	+33° 35'
Magnitude	9.9
Size	11 × 6 arcmin
Constellation	Triangulum

The barred spiral galaxy NGC 925 is relatively unknown to visual astronomers, omitted from the Messier, Caldwell, Hidden Treasures, and the Herschel 400 Catalogs. Yet this galaxy is as large and bright as the popular NGC 891, described in section "December 11: Outer Limits Galaxy." Delicate spiral arms extend outward from the central bar. Old, yellow suns dominate the central bar. Younger blue-white star clusters dominate the spiral arms, punctuated by red hydrogen clouds. NGC 925 resides at a distance of 45 million light-years in the constellation Triangulum.

Imaging. Frame NGC 925 with a small field of view similar to other comparable galaxies, to achieve good resolution. LRGB methods are almost essential, as this is a dim object. For your first try, consider binning even your luminance 2 × 2 to make sure that you have enough data to generate a quality image. Reserve this target for a calm night when you have at least several hours to devote to imaging. This would be a challenging object either from the suburbs or with a single-shot color camera.

Processing. Although routine LRGB methods work well, you will need to extract every bit of imaging data for a good result. For my image, I also created a synthetic luminance from my RGB data, and applied it as an extra luminance layer with about 25% opacity, just to boost the depth of the data. Because the object is dim, sharpen gently in the central galaxy and spiral arms to avoid generating excessive noise. If the star of magnitude 10.8, which is to the south of NGC 925, becomes bloated by your histogram stretches, apply star-shrinking techniques (see "Final Cleanup" in Chap. 15) (Fig. 12.5).

Fig. 12.5. Barred Spiral Galaxy NGC 925. East–Northeast is up.

December: Celestial Potpourri

Camera	ST10XME
Telescope	12-in. Meade LX200R at $f/7$
Field of view	23 × 15 arcmin
Exposures	Luminance clear 33 × 5 min, unbinned R 12 × 5 min, G and B each 15 × 5 min, binned 2 × 2
Scale	0.6 arcsec/pixel
Limiting magnitude	6.0

December 14: Heart Nebula and December 18: Soul Nebula

Designations	IC 1805	IC 1848
Other names	Heart Nebula	Soul or Baby
Right ascension	02 h 32.7 min	02 h 51.2 min
Declination	+61° 27'	+60° 26'
Magnitude	6.5	6.5
Size	60 × 60 arcmin	60 × 30 arcmin
Constellation	Cassiopeia	Cassiopeia

At a distance of 6,000 light-years, a cloud of energized hydrogen gas glows red in the shape of a heart (with right heart failure for you cardiologists out there). A cluster of young stars emerges at its center. A neighboring cloud of gas, the Soul Nebula is part of the same glowing cloud of hydrogen. I see a teddy bear in profile, and others claim to see a baby, but you can use your own imagination.

Fig. 12.6. (a) Heart and Soul Nebulae.

December: Celestial Potpourri

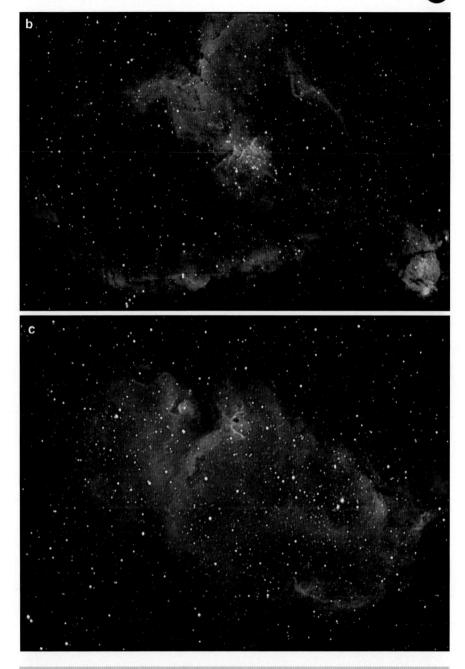

Fig. 12.6. (continued) **(b)** Heart Nebula IC 1805. East–northeast is up. **(c)** Soul Nebula IC 1848. North is up.

Imaging. You can frame both the heart and soul nebulae with a very large field of view of $3 \times 5°$, using a camera lens, as shown in Fig. 12.6(a). Or you can frame each individually, with fields of view of at least $1.5 \times 2°$. Light-pollution gradients are more pronounced with large fields. An H-alpha luminance is largely immune to light pollution, and is ideal in the suburbs. Although a red luminance is sufficient from a dark sky site for many emission nebulae, these two targets are in a rich region of the Milky Way with extensive stars that are better suppressed with an H-alpha luminance. Adjust exposure lengths to avoid blooming in many bright stars.

Processing. Both the Heart and Soul Nebulae have numerous bright stars that may require deblooming, which is best performed right after calibration. Use luminance layering to combine your images. If you use an H-alpha luminance, blend it with the red channel to maintain correct red color and to avoid dark halos around stars that project over the nebula.

Camera (both images)	ST10XME
Telescope	3.5-in. Takahashi refractor at $f/4.5$
Field of view	126×85 arcmin
Scale	7.0 arcsec/pixel
Limiting magnitude	3.5
Exposures (Heart)	Luminance Ha filter 31×5 min, binned 2×2
	R and G each 6×5 min, B 9×5 min, binned 2×2
Exposures (Soul)	Luminance Ha filter 24×5 min, binned 2×2
	R and G each 6×5 min, B 9×5 min, binned 2×2

December 18: Spiral Galaxy M77

Designation	NGC 1068
Other names	Messier 77
Right ascension	02 h 42.7 min
Declination	−00° 01′
Magnitude	8.9
Size	7 × 6 arcmin
Constellation	Cetus

All galaxies are thought to harbor black holes in their centers. Some black holes are larger than others. M77 has a massive black hole, which against intuition contributes to a bright galactic nucleus. This is generated by large quantities of material falling into the black hole. This material becomes superheated and accelerated, leading to powerful emission of radiation in the visible, radio, and X-ray frequencies. This places M77 into the category of Seyfert Galaxies, which have "active galactic nuclei." At a distance of 45 million light-years, M77 is one of the largest of Messier's galaxies. The faint outer arms extend out to a diameter of 170,000 light-years.

Imaging. Spiral galaxy M77 has outer spiral arms with low contrast, and relatively little color compared with many other spiral galaxies. Low contrast structures require a high signal-to-noise ratio to image successfully. Therefore, you should either obtain exceptionally long exposures, or bin your images to acquire data faster. Binning will decrease detail in the central portion of the galaxy, but give you richer data in less time in the pale outer arms. This image was acquired with the luminance binned 2 × 2 not only to speed data acquisition, but also because of the mediocre seeing that night, which would degrade full resolution imaging.

Processing. Routine LRGB methods work well. Use either DDP or curves to reveal the outer arms of the galaxy. Sharpening of the central bright region of the galaxy is performed routinely. The outer arms can be enhanced in Photoshop by selecting the outer arm region, deselecting the stars in this region, feathering the selection by at least 10 pixels, and then applying your sharpening tool of choice with a radius of 20 pixels. If the results seem too harsh, either fade the sharpening or perform this sharpening in a duplicate layer and apply less than 100% opacity (Fig. 12.7).

Fig. 12.7. Spiral Galaxy M77. North is up.

December: Celestial Potpourri

Camera	ST10XME
Telescope	12-inch Meade LX200R at $f/7$
Field of view	22 x 15 arcminutes
Exposures	Luminance Clear 21 x 5 min, binned 2x2 R 6 x 5 min, G 6 x 5 min, B 3 x 5 min, binned 2 x 2
Scale	1.3 arcsec/pixel
Limiting magnitude	6.0

SECTION TWO

Getting Started in CCD Imaging

CHAPTER THIRTEEN

Equipment for Astrophotography

First and Foremost, the Mount

The most important element in astrophotography is accurate tracking. You can spend tens of thousands of dollars on great optics, but 1 min of poor tracking ruins the most pristine view.

Several elements can influence tracking with any given mount. First, polar alignment helps tracking, but does not have to be much closer than 15 arcmin from the true pole. This will be treated separately in Chap. 14. Second, proper balancing of the mount helps it to work more efficiently. Most German equatorial mounts (GEM) will track even better with a slight overweighting of the east side of the mount, so that the gears are fully engaged without bouncing. Third, blur from guiding errors increases with magnification. Thus, if your mount tracks poorly, try imaging with either a smaller telescope or piggyback your camera with a shorter focal length camera lens. Most mounts, with rough polar alignment, can track a 200-mm lens for a minute or two without motion blur. Finally, if you image without autoguiding (more on that later), you can tweak your tracking with "periodic error correction" (PEC) that is built into many mounts.

Generally, amateur-level GEM track better than fork mounts. Mounts are rated by "periodic error" (PE), which is the amount that a perfectly balanced and polar aligned mount will vary during one cycle of its gears, usually between 5 and 10 min. As you might guess, the more expensive GEM mounts track best. The Paramount ME, Astro-Physics 1200, and Mountain MI 250 all have a PE of about 4 arcsec. A more moderately priced mount like the Losmandy G-11 has a PE of about 9 arcsec. Less-expensive GEMs such as the Atlas are marketed by Orion, and both Meade and Celestron have telescopes sold on GEMs. Generally you get what you pay for. Fork mounts, which come with many Schmidt–Cassegrain telescopes, typically have a higher PE than GEMS, typically between 30 and 60 arcsec. Fork mounts usually require an equatorial wedge for tracking during astrophotography. If you have a fork mount, you may choose to begin with piggyback imaging, but with extra effort a fork mount can be effective for higher resolution imaging with careful autoguiding. If you are considering the purchase of a new mount, ask an owner of the mount about how it performs for imaging. You can get excellent advice on many internet forums.

If you are preparing a budget from scratch for an astrophotography system, consider making the largest expenditure for the mount. As a rough guideline, devote 60% of your expenses for the mount, 20% to a telescope, and 20% to a camera. Most successful backyard astrophotographers have spent much more on the mount than the telescope. Eventually, for framing different objects, you will want several different telescopes with different focal lengths, so that your ultimate investment in "glass" may someday exceed your investment in "steel."

Beginner Scopes for Imaging

Among the best 100 targets in section one, about one-third were taken with "small" 3–4-in. refractors with focal lengths of 300–600 mm; one-third were taken with a "mid-size" 5.5-in. refractor with a focal length of 1,000 mm; and another one-third with "large" 8 or 12-in. Schmidt–Cassegrain telescopes at a focal length of about 2,000 mm. Different focal lengths of telescopes are useful for different angular dimensions of various celestial targets. No single focal length is effective for every object.

If you are just beginning astrophotography, I encourage you to start with a small telescope. Every aspect of astrophotography is easier with a larger field of view. Finding the object and centering it on your chip is straightforward. Tracking requirements of a mount are less critical. If you have a camera with an internal guide chip, finding a guide star is less challenging. And a quality refractor is much less expensive with a 3-in. aperture than a 5 in. or larger size.

Many high-quality refractors are now available with 3–4-in. apertures (76–100 mm) that are excellent beginner telescopes for imaging. The optics come in two basic categories: doublets and triplets. Doublets include two elements, of which one is usually made of "enhanced dispersion" (ED) glass. This reduces chromatic aberration, which can cause red, green, and blue light to come to focus separately. These ED doublets do not fully correct color, so there remains a slight focus shift between filters when doing CCD imaging. This can be corrected by refocusing after every filter change. If you are imaging with either a single-shot color CCD camera or a digital SLR, this focus shift cannot be corrected between filters, which will introduce slight blurring in your images. The same blurring will occur with a CCD camera using a clear filter for luminance. For this reason, ED doublets are often termed "semi-apochromats."

The next step in optics is the true "apochromat," which bring all visible light to the same focus. Many of these scopes use triplet optics, with one element of either fluorite or other exotic glass. Others, such as some Takahashi and TeleVue telescopes, use a four-element design. These three- and four-element apochromats typically cost twice that of similar aperture ED doublets, but can reduce or eliminate the need to refocus between filters. Furthermore, single-shot color cameras will have sharper focus. If you also use the telescope visually, the true apochromats will give the sharpest views of star clusters without false color fringing. As with mounts, you generally get what you pay for in fancy telescope optics.

When you have gained experience imaging with wide-field refractors, you may wish to move up to a larger aperture with a focal length in the range of 1,000 mm. Five- to six-inch apochromatic refractors are superb imaging scopes in this category, and some are available in moderately priced ED models and some in higher priced apochromats. Alternative telescopes in this focal length category are the Orion 190-mm $f/5.3$ Maksutov–Newtonian and the Astro-Tech 6-in. $f/9$ Ritchey–Chrétien.

Imaging with focal lengths over 1,500 mm demands precise tracking and guiding, so do not rush into this level of imaging. You can get started with an 8-in. Schmidt–Cassegrain (SCT) optical tube for $500 or less. For a little more, you can buy an improved version from Meade (now termed ACF), that reduces coma and adds a mirror lock. A mirror lock becomes essential with larger SCTs to prevent mirror flop when tracking and to allow more precise focusing. Use a focal reducer to bring your focal ratio down from $f/10$ to $f/7$ (or less) to both cut imaging time in half and reduce the demands on your mount. For the ultimate in high-resolution imaging, consider a premium Ritchey–Chrétien or other fine astrograph, but be prepared for "astronomical" prices.

Equipment for Astrophotography

Choosing a Camera

Three categories of camera are available for modern deep sky astrophotography: monochrome CCD, single-shot color CCD, and conventional digital cameras. Conventional digital cameras have a big advantage: most of us already own one. But for the best deep sky imaging, I suggest a dedicated CCD.

Both color and monochrome CCD cameras come in three categories and price points. The least expensive have no active cooling, although some of these have methods of passively dissipating heat. Because electronic noise increases with temperature, dedicated dark frames are essential (see the section "Dark Frames" in Chap. 14), but do not fully correct the images. The next category of cameras has active cooling, but no temperature regulation. These cooled cameras generate much less electronic noise, allowing longer exposures, but dark frames remain helpful, and have to be reacquired as the temperature changes. This middle category includes the Orion Sky-Shoot II and the QHY8, which use 6.0-megapixel APS-sized Sony CCD chips similar to those used in many DSLR cameras. The last category are the premium CCD cameras that have regulated cooling that can be set to a specific temperature, allowing the best control of electronic noise.

Beginning astrophotographers may prefer a single-shot color CCD camera that collects the full spectrum of color, which avoids both the fuss of filters when imaging and the chore of color channels when processing. The recent introduction of cooled CCD cameras with APS-sized color chips, priced similar to higher level DSLRs, has increased the popularity of this approach.

There is a misconception that monochrome CCD cameras are much harder to use than color cameras. My answer is that a beginner can start with a monochrome camera and stay with monochrome imaging for the first several months. A monochrome camera, when imaging without a filter, typically collects light four times as efficiently as a color chip of the same size and design. An image can have excellent depth in a fraction of the time. Furthermore, most monochrome cameras are easy to bin 2×2, 3×3, or even 9×9 to allow very short exposures of deep sky objects to check centering and framing, which can be critical when trying to locate your target on a small CCD chip. Third, your resolution is higher with a monochrome CCD chip, due to some inherent characteristics of the Bayer matrix that is commonly used for one-shot color CCD chips. Fourth, when processing your first monochrome images, you can produce satisfying pictures without the headache of color balancing and fringing.

When you feel ready to image through filters, a monochrome camera will reap additional advantages. If you have a less than perfect refractor, you can refocus separately through red, green, and blue filters to reduce most chromatic aberrations, improving the sharpness of your images. You can gather light more efficiently if you bin your red, green, and blue channels to collect color information more rapidly while acquiring luminance information at full resolution (see the section "Luminance Layering" in Chap. 15). If you image from light-polluted skies, you can try a light-pollution filter for your luminance. Narrow-band filters such as Hydrogen-alpha filters can yield excellent results from areas of severe light pollution.

Autoguiders

Nothing has popularized astrophotography more in the past decade than autoguiding. Before autoguiding, long exposure astrophotography required the astronomer to keep his or her eye glued to the eyepiece of a guide scope, making small manual adjustments to keep a guide star centered on a reticle, sometimes for an hour or more. Now, a small CCD chip keeps track of the guide star, and through a computer the autoguider gives instructions to the mount to keep the guide star centered. An autoguider helps a mediocre mount track well enough for astrophotography, and permits an excellent mount to track superbly. A good autoguider can reduce guiding errors to a fraction of an arcsecond, allowing long exposures.

Autoguiding can be divided into three categories: self-guiding cameras, off-axis guiders, and autoguiders attached to separate guide scopes. Self-guiding cameras use elements internal to the camera to provide autoguiding and come in two types. One type, used by Santa Barbara Imaging Group (SBIG), has a second internal guide chip within the camera. It uses light next to the main imaging chip to guide the mount. The advantages of this design are as follows. (1) Allow both the imaging chip and guiding chip to be focused simultaneously, (2) prevent differential flexure in a guide scope from causing tracking errors, and (3) allow the guiding chip to track during slow mirror shift or sag in a SCT. The second type, used by some Starlight Express cameras, uses alternate rows of the main imaging chip for guiding and imaging. The additional advantage of this design is lower cost, but because half of the rows are guiding at any time, the total imaging time doubles. Both types of self-guiding cameras have difficulty with narrow-band imaging, because the guiding takes place behind the filters, dramatically dimming the guide star.

Off-axis guiding uses an extension in front of the camera that contains a small pick-off mirror just outside of the field of the camera. The mirror diverts light to a separate guiding camera, and allows guiding during the exposure. Like self-guiding cameras, off-axis guiders are immune to both flexure problems in a guide scope and slow mirror shift in a SCT. Because the off-axis guider uses light in front of the filter wheel, it is not hindered by narrow-band filters. If you are considering an off-axis guider, make sure that the longer imaging train is sturdy enough to avoid any additional flexure, and that you have enough back focus. A recent CCD camera design by Quantum Scientific Imaging (QSI) incorporates an integral off-axis guider in front of the internal filter wheel that adds only a ½ in. of backfocus.

The third category of autoguiding uses a separate guidescope. This technique can use a focal length for the guide scope as short as one-fourth that of the main scope, easing the location of guide stars. Narrow-band imaging through the main scope does not affect the guide scope. This method works best for guiding one refractor with another, when the two scopes are rigidly connected. This does not work as well for large reflectors or SCTs, in which slight mirror sag or slow shift of the primary mirror can cause blurring of the image that is undetected by the guider.

Investing Wisely in Software

Digital imaging requires software for both controlling the camera and processing the images. Many CCD manufacturers include basic software for operating their cameras, which may be sufficient for acquiring your data. When you are ready to consider a more sophisticated and versatile program, check manufacturer Web sites for free trials to make sure that you will be satisfied with the investment. As of this writing, you can download trials of image processing programs like CCDstack or integrated camera control and processing programs like Maxim DL, Astroart, and Nebulosity. Sometimes these and other programs are discounted at astronomy shows.

After basic calibration and combining, many astrophotographers transfer their images into Photoshop or other similar programs for further processing. Unfortunately, the latest version of Adobe Photoshop can cost as much as a small apochromatic telescope. You can choose another less expensive program, including Adobe's own ImageReady or Photoshop Elements, but the full Photoshop program has the advantage of being the de facto standard for final processing of celestial images. Many astronomers (including me) have legally purchased discontinued versions of Photoshop that have never been opened or registered, from EBay or discount vendors, at a fraction of the cost of the latest version. Most of the images in this book were processed with Photoshop 5.5, which I have recently upgraded to version CS2. If you use Photoshop versions 6 or earlier, you will be limited to processing in 8-bit format, which reduces dynamic range. For these early versions of Photoshop, consider downloading either freeware FITS Liberator (http://www.spacetelescope.org/projects/fits_liberator/) or shareware FITS Plugin (http://astroshed.com/software.html) to allow Photoshop to apply curves and levels to 16-bit FITS images.

CHAPTER FOURTEEN

Acquiring the Image

Siting the Telescope

Whether you image from your backyard or from a remote dark sky site, you should try to simplify your preparation. If you have a portable setup, you must transport your equipment, assemble the mount, attach your telescope, connect power cables, polar align, and attach your camera, before you even begin to search for your target. Should inclement weather appear, each step may need to be reversed hastily, adding to your frustration. A permanent setup avoids these inconveniences. If you live in a rural or suburban area, this might be an option at your home. The simplest arrangement is a permanent pier, either on a deck or in a yard. A pier will be more stable than a tripod, already leveled, and can be oriented or marked to speed polar alignment. You can run underground conduits to your pier containing electrical supply and cables for remote control on cold nights.

Once you have a pier, consider constructing a small roll-off shed just large enough to cover up your telescope. Details of my astro-shed were published in the November 2004 issue of *Astronomy* magazine. My shed measures 6-ft. high, 4-ft. wide, and 3-ft. deep, and rolls snugly around a pier anchored beneath my backyard deck. I can keep my CCD setup protected from the elements, yet begin or shut down an imaging session in less than 10 min (see Fig. 14.1).

Ultimately, an observatory provides an ideal setting for astrophotography, if permitted by local building codes, your budget, and your spouse! You can either build one yourself, pay someone else to build it for you, or buy a prefabricated fiberglass dome. Should you decide to build it yourself, you can refer to several books on the subject, and gain additional advice on Yahoo discussion groups. Most homemade observatories use a roll-off roof. You can buy construction plans and accessories like tracks, rollers, and piers from companies like Backyard Observatories and Sky Shed. If, like me, you lack carpentry skills but prefer a roll off design for esthetic reasons, you can get help. Both Backyard Observatories and Sky Shed sell prefabricated observatory kits, and they may be willing to travel to your site and complete the construction. Because they have assembled many observatories, they can offer quality construction at reasonable prices (see Fig. 14.2).

Prefabricated fiberglass or metal domes can be a better choice if you have frequent wind or nearby light pollution, both of which are better shielded by domes. Be careful of translucent fiberglass designs that allow light to penetrate the observatory, which can create a greenhouse effect, possibly damaging your telescope in the summer and slowing temperature equilibration year round. Some sophisticated domes will rotate to track along with your telescope. Clamshell designs such as AstroHaven can be aligned east-to-west to allow tracking through the opening without the need for dome rotation.

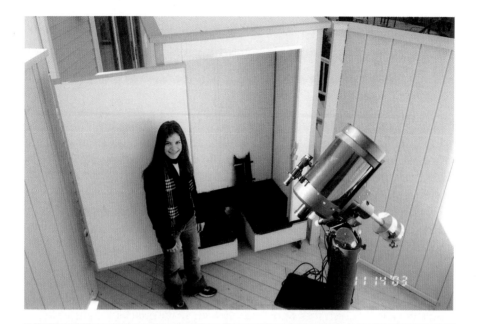

Fig. 14.1. The author's daughter stands next to their Astroshed, which is mounted on wheels, and can roll forward to protect their 8-in. SCT telescope mounted on a pier.

Fig. 14.2. Backyard observatories built this roll-off roof observatory for the author in 4 days. The open area contains two piers. A small warm room is on the right.

Polar Alignment

Polar alignment is important to prevent drift during astronomical exposures. Accurate alignment is essential for pointing accuracy of "Go To" telescopes, so that your target appears on your small CCD chip. Many astronomers obsess about achieving precise polar alignment before imaging. While I agree that polar alignment is important, your need for accuracy depends largely on your image scale. For imaging at shorter focal lengths, aligning to Polaris may be sufficient for exposures of 3 min or less, especially if you are autoguiding your exposures. Just aligning on Polaris will get you within 1°. If you have a German equatorial mount with a sight hole along the polar axis, you can perform this rough alignment before mounting your telescope. First level your mount, then adjust your mount to your latitude, and finally center Polaris through this sight hole. If your mount lacks a sight hole, or is a fork mount on an equatorial wedge, you can use your finder instead. For this approach, your finder should already be aligned to your scope, and your setting circles should be reasonably accurate. First level your telescope, then move your setting circles to the coordinates of Polaris (about Dec 89.3° and RA 2 h 30 min), and then adjust your altitude and azimuth to center Polaris in the finder.

For longer focal lengths or longer exposures, more accuracy is needed. My recommendation is to obtain a polar alignment scope (PAS), if one is available for your mount. When first using the PAS, align it to your polar axis by pointing at a distant terrestrial object, and using the PAS calibration screws to keep the target centered as you rotate the RA axis 180°. At night, first level your mount, set your latitude, and point your RA axis toward Polaris. Begin by locating Polaris through the PAS. Most polar scopes have a reticle indicating the Big Dipper on one side and Cassiopeia on the other side, to help define the direction to offset Polaris from the pole. This correction should get you to within a quarter of a degree. If your PAS reticle has a marker for Delta Ursa Minor, and your skies are dark enough to see this star, then move the mount to place both Polaris and Delta on their reticle markings. This can give you accuracy to 5 arcmin from the pole. Most of the images in this book were obtained using just a PAS for polar alignment.

Many Go To mounts with digital setting circles also include a polar alignment routine in their imbedded software. This software may allow accuracy comparable to a PAS. The precision may be enhanced by including more calibration stars. This accuracy may be sufficient for most astrophotography applications.

Acquiring the Image

The best polar alignment can be achieved with drift alignment methods. Drift alignment was more critical in the years before CCD cameras, when film photographers took exposures of an hour or more. Because drift alignment can take up to an hour, you may only want to attempt this if you have a permanent setup on a pier in an astroshed or observatory. If the steps of drift alignment seem too complicated to you, consider Pempro from CCDWare, which has a relatively simplified routine for drift alignment using your CCD camera.

Choosing the Target

A successful imaging session begins well before darkness, when you choose your target. You should plan to image targets that are (1) high in the sky during your session, (2) appropriate in size to the field of view (FOV) produced by your camera and telescope, (3) accessible to the sensitivity of your camera and telescope, (4) within the range of objects that local light pollution allows, and (5) within your technical expertise as an astrophotographer.

The best imaging occurs at the celestial object's highest altitude (which occurs when it crosses the meridian) for four reasons. First, at the meridian, the object shows the greatest freedom from atmospheric seeing, because the camera looks through less turbulent air. Second, the object is also at its brightest because of "atmospheric extinction", which represents the absorption and scattering of light by dust, water, and air pollution. A magnitude 8.0 galaxy at the zenith drops to 8.12 at an altitude of 45° and 9.0 at 12.5°. Blue light is dimmed more severely than red light, because dust scatters blue light more effectively. Third, the atmosphere bends light to a greater degree closer to the horizon. This "atmospheric refraction" functions like a prism because refraction varies for different wavelengths (colors) of light. For example, red and blue light from the same star will be stretched several arcseconds apart below 30° altitude. Finally, light pollution and resulting background gradients increase dramatically at lower altitudes. The "Best Targets" are arranged by the date that an object is on the meridian at 9 p.m. standard time in the center of your time zone. Plan to image a given object when it is near the meridian. If you are imaging at a different time, adjust the date to 1 month earlier for every 2 h later.

The best images frame the target. Avoid objects that are too large for the FOV created by your camera and telescope (see Table 14.1), that make the object appear crowded. Similarly, avoid objects that will appear too small and lost in the image. The "Best Targets" provide the sizes of the objects and recommendations for FOVs. Often several different FOVs are suggested for framing different aspects of an object, and fully a quarter of the Best Targets can be framed to include a second or third object. You can enhance the versatility of your telescope by using a reducer to increase your FOV or a Barlow lens to decrease your FOV.

Some objects are very faint, and thus demand highly efficient CCD detectors, relatively fast focal ratios, and long exposure times. The faint targets may not be rewarding to the user of a less-efficient one-shot color

Acquiring the Image

chip or DSLR, or for those with either slow-focal-ratio telescopes or imprecise mounts. These same faint objects are difficult to image from heavily light-polluted areas, where sky glow may overwhelm nebulosity. An IDAS light-pollution filter can help suppress moderate light pollution when imaging the brighter objects. Narrow-band filters can work with efficient CCD detectors, fast focal ratios, and very long exposures to yield interesting results amid severe light pollution.

As you begin, concentrate on the larger, brighter, and familiar objects that can be imaged most easily. With experience, the entire list should become accessible.

Calculating Your Fields of View and Scale

Table 14.1 shows the FOV and image scale for several popular imaging chips when used through telescopes of various focal lengths. Not all focal lengths of telescopes nor sizes of chips could be included on the chart. If your scope or your chip is not included, either pick a close combination, or perform your own calculation within software such as TheSky. If you are using a digital SLR, you probably can use the fourth column, which is the Sony chip used in the Orion StarShoot Pro, but is also the APS size used in many digital SLR cameras.

All of the Best Targets can be imaged successfully using any of these imaging chips with the appropriate focal length for framing. Even a small CCD chip, when combined with a 200-mm telephoto lens, can obtain a large enough FOV to frame large objects such as the North American Nebula and the Witch Head Nebula. Larger chips allow framing the same objects with longer focal lengths, permitting higher resolution, but this can introduce other challenges.

Resolution in an image is limited by atmospheric seeing, tracking accuracy, focus precision, and image scale. Under ideal circumstances, the greatest sharpness can be achieved at an image scale of about one-third the atmospheric seeing, assuming that you obtain enough image data to overcome other sources of noise. Thus, if your skies average 3 arcsec seeing, you will not gain any more resolution by imaging below 1 arcsec/pixel. If your scale is much less than that, then you would get better results by binning your pixels 2×2 to gather light faster and improve your signal-to-noise ratio. Depending on your mount, telescope, and experience, tracking errors and imperfect focus may exceed the distortion of atmospheric seeing, and therefore imaging above 2 arcsec/pixel may be a more efficient use of imaging time. Pleasing results are possible at larger image scales, as evidenced by many of the images in this book.

Acquiring the Image

Table 14.1. Popular combinations of imaging chips and focal lengths, and resulting fields of view (fov) in arcminutes and scale in arcseconds/pixel.

Imaging chip	Kodak KAF-0402			Kodak KAI-2020			Kodak KAF-3200			Sony ICX413AQ			Kodak KAI-11002		
Array	765×510			1,600×1,200			2,184×1,472			3,032×2,016			4,008×2,672		
Pixel size (µm)	9			7.4			6.8			7.8			9		
CCD size (mm)	7×5			12×9			15×10			23×15			36×25		
Focal length (mm)	FOV	Scale		FOV	Scale		FOV	Scale		FOV	Scale		FOV	Scale	
200	118×79	9.3		203×153	7.6		255×172	7		390×259	7.8		617×429	9.3	
300	79×53	6.2		136×102	5.1		170×115	4.7		260×173	5.2		412×286	6.2	
500	47×32	3.7		81×61	3.1		102×69	2.8		156×104	3.1		247×172	3.7	
700	34×23	2.7		58×44	2.2		73×49	2		111×74	2.2		177×123	2.7	
1,000	24×16	1.9		41×31	1.5		51×34	1.4		78×52	1.6		124×86	1.9	
1,500	16×11	1.2		27×20	1		34×23	0.9		52×35	1		83×57	1.2	
2,000	12×8	0.9		20×15	0.8		26×17	0.7		39×26	0.8		62×43	0.9	
2,500	9×6	0.7		16×12	0.6		20×14	0.6		31×21	0.6		50×34	0.7	

Finding, Centering, and Framing the Target

Visually, you would have no problem finding objects with a 12-in. SCT. You would place a wide-field eyepiece into the diagonal, press GoTo on your mount or star hop with your finder, and the object would appear somewhere in your field. But if you try the same exercise with a CCD camera attached to the telescope, the object sometimes is not on the chip. Why? With this telescope, a wide-field 40-mm eyepiece covers a FOV of 54 arcmin round, whereas an average CCD camera like the ST2000 covers a rectangle 10×14 arcmin. The difference in area is 16 times larger with this eyepiece than with this camera. Even with a fixed pier, excellent polar alignment, and a premium mount, you may still have difficulty locating some celestial objects. One option, if you have a GoTo mount, is to buy a telescope pointing program like TPoint or MaxPoint to increase the accuracy of your pointing. Another option, which I use, is to "star hop" to your target. Before slewing to your object, first slew to a bright star roughly half way to your target. Take a preliminary "focus" image through the camera. If the star is visible, center the star and then recalibrate or synchronize your GoTo mount to this star location. This may be a good time to confirm that your finder is aligned to your mount. For large SCTs, mirror flop may change the location enough so that your finder no longer is aligned to the main telescope. Your next slew, to your target, is more likely to show your object on the chip. If needed, you may perform intermediate slews and recalibrate or synchronize to ultimately find your target.

If you do not have a GoTo mount, you may still be successful at finding your target if you are expert at star hopping visually with a finder, especially if you are using a short focal length telescope or a large imaging chip, or both. An alternative is to create and modify a wide-field eyepiece to make it parfocal with your CCD camera. Many astronomical accessory companies market parfocal rings that attach to the barrel of your existing eyepieces. Simply exchange your eyepiece with your CCD camera after you have centered the object visually. A third option is to have a flip mirror between your camera and the telescope, with an eyepiece in the side. When the mirror is up, it reflects light from the telescope to the eyepiece, which can be made parfocal with parfocal rings. Once centered and focused, the mirror can be lowered to allow light to reach directly to the camera. The disadvantages of a flip mirror are the increase in backfocus, and the additional flexure that may be created in a long imaging train.

Acquiring the Image

Once your object is visible, your first instinct is to center the object. If you use an internal guider like the SBIG cameras, you may need to rotate your camera to place a suitable guide star on the chip. Because your field is rectangular and because most celestial targets are not round, you may choose to turn your camera to place the long axis of your target along the wider portion of your chip.

Now you must consider composition. An oval object perfectly aligned to the chip, or a round object in the precise center of the chip, may look static. The oval object will look better at an angle from corner-to-opposite-corner. A round object will usually appear more interesting slightly off center. Try to find some secondary object to balance the main target. Ideally this will be another interesting object, as can be found in a quarter of the Best Targets. Other times, this will be a small background galaxy or a faint region of nearby nebulosity. A third choice would be a bright star or group of stars. These balancing objects help establish an environment and sense of perspective for your primary target. To check your composition with a CCD camera, obtain some brief exposures with luminance and 3×3 or even 9×9 binning. Observe how the nebulosity, stars, or galaxies fill the image, and adjust your framing accordingly. When you are satisfied with your composition, recalibrate or synchronize again, so that you can return quickly to this frame.

Focusing

Accurate focus is essential, and benefits from a good focuser. Most SCTs are difficult to focus by using the main focus control that moves the primary mirror. For these scopes, consider purchasing a dual speed Crayford focuser for the visual back, and locking the main focus knob if possible. Manual focus is easier to fine tune with a dual speed focuser, or an electric dual speed focuser. If you are purchasing a new refractor, look for a dual speed focuser that is also sturdy enough to support a CCD or digital camera.

Speed up manual focus first with a rough focus based on prior experience with the focuser. If you have a position scale on your focuser, write down the position of good focus, perhaps on a piece of masking tape attached to the telescope. If you are using a crayford or rack-and-pinion type focuser without a position scale, tie a plastic ruler to your scope, measure the location of good focus, and mark it on the ruler for your next session. If you have already adapted an eyepiece to be parfocal with your camera, you can use the eyepiece to center your target and provide a rough focus. A flip mirror system with a parfocal eyepiece can also help to both center the target and provide a rough focus (see the section "Finding, Centering, and Framing the Target").

Fine focus is done through the CCD camera. For most methods, first center a reasonably bright star. Avoid the brightest stars that may saturate the image. Select a small subframe around the star for focusing to speed up downloading during the focus session. Many camera control programs include routines to aid manual focus, using parameters such as maximum pixel value, full-width-half-maximum, or sharpness algorithms. These programs are time consuming and can be error prone due to atmospheric seeing, and therefore may be best reserved for remotely focusing the camera with an electronic focuser.

If you are standing next to your telescope, you can focus faster and just as accurately with an aperture mask. The most commonly used types of aperture masks use either circles or triangles cut out of dew caps (see Fig. 14.3). You can buy a sturdy mask that will fit your telescope, with circular holes from Kendrick or with triangular holes from ScopeStuff. To focus, simply place the mask over the front of the telescope, and adjust focus until three star images merge into one. You can make an even better mask yourself with two triangles from a plastic bucket or cup, matching the bottom of the bucket to the outer diameter of the telescope aperture. Cut inverted triangles at the base of the bucket using a razor knife. Add adhesive felt to the inside of the base of the bucket until it fits snuggly over the telescope. As you approach focus on a bright star, the two triangles begin to merge. At perfect focus, thin diffraction spikes appear at the edge of the star. The better the focus, the thinner and brighter the spikes become. Magnify

Acquiring the Image

Fig. 14.3. Three types of aperture masks, made by ScopeStuff (*left*), by Kendrick (*bottom right*), and by the author from a plastic bucket (*upper right*).

the image on your computer screen to see the spikes clearly. Do not forget to remove the aperture mask after focusing!

Many electronic focusers can work with @Focus or FocusMax to provide the most accurate focus. @Focus comes with CCDSoft whereas FocusMax is a free download. Both use multiple automated samples of the approach to focus to define the center of the focus range. Focus achieved in this way can stay sharp during minor temperature changes. FocusMax requires more setup time to get started, now using a programming "wizard" to simplify the process, but once programed for your imaging system it finds focus more rapidly than @Focus. Both are very accurate.

Focus is not stable through the night. Temperature changes focus, so consider refocusing at least every hour. SCT mirror sag and flop can also alter focus, especially after meridian swaps, so always refocus your SCT after a meridian flip. If you change filters, consider refocusing, because even "parfocal" filters often do not come to the same exact focus. If using a semi-apochromatic refractor or camera lens, always expect some focus shift between filters. Atmospheric refraction will blur images taken without filters when the target is at lower altitudes, so obtain your single-shot color or unfiltered luminance images when your target is near the meridian to minimize this effect (see the section "Choosing the Target").

Autoguiding

Regardless of the type of autoguider, you must first find a guide star. If using an off-axis guider or an SBIG self-guided camera, you may need to rotate the guider or the camera until a bright enough star for guiding falls onto the chip. A mapping program like The Sky may help you find a guide star in advance.

An ideal guide exposure is about 5 s. If the guide star is so bright that it is saturating, decrease the guide exposure or find another star. Avoid guide exposures less than 2 s, because the guider may start chasing atmospheric turbulence that can shift the apparent location of a star by an arcsecond or more. If the best available guide star is dim, you can increase guide exposures up to 30 s to reach an ADU of at least 2000, to prevent the camera from confusing the guide star with electronic noise. If using Maxim DL, make sure to enable Auto Dark in the Focus Tab, which is automatic in CCDSoft.

You must calibrate the autoguider to your telescope, mount, and declination before imaging an object. Most camera control programs perform this calibration after a few inputs from you. First, place the star at the center of the guide chip when beginning. Make sure that you are using the brightest star on the chip, because the program will track the brightest star. Avoid calibrating when there is a second star in the field as bright as your guide star, in which case you may need to shift your guiding field for calibration. You can calibrate on a star up to a few degrees away from your ultimate target, and still have results accurate enough for autoguiding.

You will need to enter a calibration time for the X- and Y-axis. The shorter your focal length, the longer your calibration times will need to be, to allow the star to move across enough of the guider chip. If you are guiding at 0.5× sidereal, try setting a calibration time of 20 s when guiding at 500 mm or 10 s at 1,500 mm. After calibration, check the calibration results to confirm that the X- and Y axis are perpendicular to each other, and that the plus and minus portions of the graph are roughly symmetrical.

Additional important settings are aggressiveness and minimum/maximum move. I suggest beginning with aggressiveness set to half of the maximum, which means that the mount will try to correct half of the guiding error with each correction. This prevents overcorrection, which can have the mount bouncing back and forth. The minimum move should be set high enough to prevent the mount from initiating a correction for tiny tracking errors, and any setting from 0.01 to 0.1 s may be appropriate depending on your mount. The maximum move for a good

Acquiring the Image

mount should be 1 s. For a less accurate mount, consider 2 s. If you allow more than these times, a cosmic ray hit on the edge of your guide chip may ruin your entire exposure.

When moving on to another target, you should recalibrate your autoguider. Most programs allow you to enter a declination when calibrating, promising to scale this result to other declinations. Do not believe it. Recalibrating takes only a minute or two, which is only a fraction of the time lost with a single-blurred exposure.

Exposing

Most astronomical imaging combines multiple exposures of shorter lengths. The main reasons are the limitations of the camera and the mount. Digital SLR cameras and inexpensive uncooled CCD cameras are usually limited to exposures of a minute or less because of the thermal noise in the image. Dozens or hundreds of images may be combined.

For cooled CCD cameras, longer exposures are possible, but shorter exposures can be stacked to mask tracking errors by the mount. You should perform some test exposures to judge how long of an exposure you can take before tracking errors become objectionable. Shorter focal lengths permit longer exposures before tracking errors appear. Five 1-min exposures summed together have the same signal as a single 5-min exposure, but the noise is higher in the summed image, so that the ultimate signal-to-noise is better in the single, long exposure. However, longer individual exposures can become overwhelmed by light pollution. Also, blooming in stars becomes worse with longer exposures, especially in non-antiblooming cameras.

For imaging of most celestial objects with a cooled CCD camera, 5-min individual exposures provide a good compromise between image quality, control of sky glow, control of blooming, and tracking accuracy. Ten-minute exposures may work better with many antiblooming cameras and single-shot color cameras, which often have lower sensitivity to light. These times can be doubled for narrow-band imaging, because both light pollution and star blooming will be suppressed, assuming your mount and autoguider track precisely. Shorter exposures are essential for brighter objects, like the Pleiades or the Orion Nebula, where even 1-min exposures lead to blooming. Some bright open clusters, like the double cluster, may require an intermediate exposure of a few minutes.

Although imaging of very dim objects from a dark sky site may benefit from exposures over 20 min if there are no bright field stars, such long exposures have serious drawbacks. First, if a passing cloud blocks the guide star for a minute, or your neighbor's car shines its lights directly into your telescope, the entire exposure gets ruined. The same event would only ruin one of the four 5-min exposures. Second, if you are doing RGB imaging, you need to get exposures in each channel. Ideally, you should have at least three images in each channel to be able to eliminate artifacts from cosmic ray hits and passing airplanes. If you are obtaining 20-min exposures, you then must have over 3 h of imaging to cover all the three color channels. For these reasons, most of the 100 Best Targets were obtained with 5-min subexposures.

Acquiring the Image

Some final pointers are as follows. (1) For monochrome cameras with filters, try to obtain your luminance and blue exposures near the meridian, where the effects of atmospheric refraction and extinction are less severe. (2) Make sure that your software is set to autosave your images. (3) If you use subframes for focusing, make sure to return to full frame before imaging, and remove any aperture mask!

Dark Frames

After you have tackled focusing and tracking, the next task is to remove image noise. The largest contributor to noise in an image is thermal noise, also called dark current. Most of this noise can be effectively removed by the use of a dark frame, which combines the dark current with the read noise.

A dark frame is obtained with an exposure both at the same temperature and of the same duration as your light exposure. If your CCD camera does not have a shutter, you can obtain a dark frame by putting the dew cap on your telescope when you take the exposure.

Thermal noise increases with the length of the exposure and decreases with lower temperature. A cooled CCD camera will reduce the thermal noise on your light images, so that your dark frames would not have to work as hard.

Just as with a light exposure, a dark frame will have some random noise (shot noise), and also may be affected by cosmic ray impacts. These problems can be remedied by using multiple dark frame exposures, and then combining them (see the section "Combining Images" in Chap. 15). At the very least, try to obtain at least three dark frames for every combination of exposure length, temperature, and binning. If you are imaging bright objects with a camera cooled to −20°C or lower, little may be gained by obtaining more than five dark frames. However, if you are imaging a faint nebula that is barely emerging from background noise, or if your camera cannot be cooled below −10°C, or you are obtaining very long exposures through narrow-band filters, your image will likely benefit from combining more dark frames.

The best dark frames are obtained the same night as the light exposures. The most convenient time may be after you have finished your imaging for the night, before you warm up your camera. You can just turn off your mount, put the dew cap on your telescope, and set your camera to take a number of dark frames while you take a long deserved nap. You can continue to use the same dark frames for images taken on other nights, as long as the exposure time, bin mode, and temperature settings on the chip are the same. Over the course of several days and weeks, the pattern of the dark frame will change slightly, so that most imagers will take a new set of dark frames each month.

If your dark frames have a different exposure time than your light exposures, you can scale the dark frame. Scaling dark frames requires that you also obtain a bias frame. A bias frame is like a dark frame with an

Acquiring the Image

exposure of 0 s. Thus, the bias frame measures primarily the read noise. Astronomical software can subtract the bias frame from the dark frame, scale the dark current to the length of exposure (a linear relationship), and add back the bias frame. For example, you could take 10-min dark frames and bias frames at the same temperature, and use scaling to create dark frames for both your 10-min H-alpha and your 5-min RGB filters. In practice, scaling is imprecise and adds noise, and should only be used if you are unable to obtain dark frames with the same exposures as your light frames.

Flat Fields

Flat fields allow correction for imperfections in the entire optical system. These flaws include vignetting, dust on filters or other glass surfaces, and non-uniformity in the sensitivity of the camera's pixels. A flat field is an image of a uniform light source. Image processing software applies this flat field to the light exposure, which not only removes vignetting and dust shadows but also corrects for differences in sensitivity of pixels.

Several popular methods of obtaining flat fields include sky flats, t-shirt flats, dome flats, and light box flats. Sky flats are typically obtained at dusk or dawn, pointing far away from the sun, and avoiding daylight that may overexpose the CCD chip. Sky flats require no other equipment and are popular with remote or automated imaging setups. Sky flats have two disadvantages. First, sky flats introduce gradients into the image, which have to be corrected during final processing. Second, sky flats will include many stars. Thus, at least three sky flats should be obtained, with either the mount tracking turned off to streak the stars or with slight movement of the telescope to shift star position between exposures; these flats can then be combined with a sigma combine or median combine to subtract the stars.

A T-shirt flat uses a white t-shirt or sheet placed over the telescope aperture, held taut by rubber bands or bungee cords. The white cloth is an effective diffuser. To avoid gradients on the surface of the cloth, consider obtaining your t-shirt flats at dusk or dawn, and pointing your telescope toward the sky but away from the sun. Most of the images in this book utilized t-shirt flats at dusk or dawn.

Light boxes can be constructed or purchased to fit over your telescope and provide uniform illumination. The light source should be sufficiently diffused to avoid introduction of gradients. If you are using a single-shot color CCD or a DSLR, be careful of introducing color bias with a light box. Most of the images in this book obtained through my 5.5-in. refractor utilized a light box for flat fields.

All flat fields should be obtained with the same optical system as light exposures. If you are using multiple filters, the best results are obtained with a different flat field for each filter, because each filter may have a unique pattern of dust spots and other optical aberrations. Ideally, flat-field exposures should be obtained the same night (or twilight) as your regular exposures, because dust spots can change in position, but some astronomers only change their flat fields every month. The exposure time should target about half of saturation point of the camera to maximize signal-to-noise. This requires close attention during twilight as light changes rapidly.

Acquiring the Image

Some astrophotographers suggest obtaining dark frames for flat fields. However, because flat-field exposures are usually only a few seconds, and dark current is minimal on these short exposures, you can just get bias frames at the same temperature as the flat fields, and subtract the bias from the flat. Maxim DL allows subtracting bias from flats automatically, but other programs may require you to identify the bias as a "dark for flat."

CHAPTER FIFTEEN

The Order of Image Processing

Image Reduction/Calibration

The first step in processing your exposures should be the dark frame subtraction and flat-field correction. Some programs, like CCDSoft, call this process image reduction, whereas other programs, like Maxim DL, call this image calibration. Most image processing programs have a routine to perform this efficiently on all of your images. Make sure that the dark frames match the temperature, the exposure time, and the bin mode of your light exposures. Also, confirm that your flat fields match the optical system of your light exposures. See the section on "Combining Images" for suggestions on dark frame and flat-field combining.

Your calibrated images should be saved in a different folder from your raw images. This way, if your images are not calibrated correctly, you can repeat the process on your raw images. The calibrated image should be virtually free of the scattered hot pixels that plague the raw image. Vignetting and dust shadows should be improved substantially.

The Order of Image Processing

Optional Steps

Deblooming software is included with many image processing programs, such as Maxim DL, Astroart 4.0, CCDStack, and can be purchased from NewAstro as a plug-in for CCDSoft. Deblooming should be performed on calibrated images before any alignment or combining. Alignment resamples the image data, which can blur the borders of the bloom and reduce the effectiveness of deblooming programs. If your image has only one or two small blooms, you may choose to skip this step altogether, because the deblooming program may adversely affect bright areas of the image. A few small blooms may be easily removed with final processing. If you have multiple or large blooms, deblooming software can correct most of the bloom, leaving small residual artifacts that are more easily repaired during final retouching.

If you obtained your color channels binned differently from your luminance channel, this is a good time to resize your binned images to the same size as your luminance. This allows you to align all of your images together, at the greater precision of full resolution.

Some cameras will have scattered bright and dark pixels even after proper calibration with dark frames and flat fields. If these appear excessive, they can be removed with your astronomical imaging software. However, these pixels will usually diminish or vanish after combining images, especially if a median combine or sigma combine approach is used.

Aligning Images

After your individual exposures are calibrated, debloomed (if needed), and resized to the same size, you are ready to align and combine them. Some processing programs allow you to perform both alignment and combining as a single step, but that places complete trust in the accuracy of the alignment process. A better approach, at least until you develop confidence in the alignment process, is to separate the two processes.

Different programs perform alignment in different ways, and with various degrees of complexity. CCDSoft does batch alignment on a folder of images by matching star patterns and is usually quite accurate, but can be fooled by areas of blooming or bright cosmic ray hits. Maxim DL has many options for alignment, but the most reliable of these is two-star alignment. Two-star alignment will correct for position and rotation, but not differences in scale. Maxim's auto-star-matching routine will correct for differences in scale, and thus is very useful for aligning images from different telescopes, but auto-star-matching will often fail to accurately align all of a larger group of images. CCDStack has very accurate matching routines and can correct for position, rotation, and scale, but this program can be more difficult to master. Most other programs have variations of these alignment routines. Always inspect your images after alignment to confirm that the process was successful, before proceeding to combining your images.

Every alignment routine resamples the images, which introduces a slight degree of blur. Therefore, try to limit your alignment to a single run of the alignment program. If your alignment was poor, return to the original images before modifying your alignment procedure. Do not subject the same images to multiple alignment steps, at the risk of diminishing your resolution.

The Order of Image Processing

Combining Images

If you are using filtered images, you will need to combine each filter channel separately. Your aligned images can be combined by either addition, average, median combine, hybrid average/median, sigma combine, or standard deviation masking. Addition and average provide the highest signal-to-noise ratios, because all of the data is used to create the image. Average avoids the risk of creating pixels with a higher value than saturation. Addition may be better than averaging for very dim images, such as those obtained with narrow-band filters. Both addition and average allow cosmic ray hits and airplane trails to remain as dim artifacts on your final image.

Median combine examines all of the images, and displays just the median value for each pixel. This eliminates cosmic ray and airplane artifacts. However, median combine has a lower signal-to-noise ratio than averaging, because only one data point is ultimately used for each pixel.

You can use a hybrid method of average and median technique to try to blend the benefits of both. This method requires at least six images to combine, but can work with many more. The images are divided into subgroups of at least three images. Each subgroup is median combined, which eliminates cosmic rays and airplane trails. Then, the median combined images from the subgroups are averaged. In principle, this achieves the artifact suppression of median combine, with some of the benefits of increased signal to noise provided by averaging. This technique has been replaced by the more efficient routines of sigma combine and standard deviation masking, but the hybrid method remains an option if your software only allows median combine and averaging.

Sigma combine (also called sigma clip or sigma reject) and standard deviation masking are similar routines that provide both the artifact rejection of median combine, and the improved signal-to-noise of averaging. Both require at least five images for good results. Most image processing programs now include these routines. If your software does not, you can download for free a sigma program written by Ray Gralak at http://www.gralak.com/Sigma/. With all of these programs, combine your light images with normalization at about 60%, check to ignore the black border caused by alignment, and enter a sigma value of 0.5–1.0.

Sigma combine works well not only on light exposures, but also on dark frames, bias frames, and flat fields. For dark frames, do not use normalization and do not check to ignore the black border. For flat frames, use normalization at about 60% but do not check to ignore the black border. If you use the calibration routines in Maxim DL, make sure that your dark, bias, and flat frames are each assigned to use the proper combine settings before they become applied to your light images.

Deconvolution

Once your light images are combined, you can try deconvolution to reduce blurring caused by atmospheric instability, poor tracking, or imperfect focus. This method works best for images that are both high resolution and have a high signal-to-noise ratio. Different astronomical imaging programs have different implementations of deconvolution, and will vary on input variables such as iterations, noise models, and point spread functions. If your program either lacks deconvolution, or the process seems too complex, consider downloading the free program CCDSharp from the SBIG Web site. CCDSharp is easy to use and works well on most images with just five iterations (or six iterations for small bright planetary nebulae).

Deconvolution works best on the luminance channel of an LRGB image (see the section "Luminance Layering"). On routine RGB images, you can perform deconvolution of your individual color channels, but rims of color may appear around stars after color combining. This artifact, if it appears, can be fixed in final processing by slightly blurring the color component of your image (in Photoshop: Filter > Blur > Gaussian Blur (radius 1.0 pixel) > OK; then Edit > Fade Gaussian Blur, change mode to Color, and adjust Opacity to less than 100% until color is blurred just enough to hide the artifact).

The Order of Image Processing

Color Combining

Once you have combined images for your red, green, and blue channels, you are ready to color combine. These channels should already be aligned to each other, but can be realigned if necessary. You can choose to combine the color channels in your astronomy software or in an imaging processing software like Photoshop.

Astronomy software will allow you to normalize backgrounds, which is usually a good idea to neutralize the background. Astronomy software often allows you to weight each color channel. Some cameras have poor blue sensitivity whereas others have poor red sensitivity. You can adjust the weighting of the channels to achieve a better white balance in your image.

If you choose to combine color in Photoshop, each image must first be converted to .tiff format. Then, open all the three images in Photoshop, choose Merge Channels under the Channels dialog, then choose RGB Color and number of channels = 3. Do not be surprised if your images first appear almost black in Photoshop. Whereas astronomy programs have histogram viewing features that allows seeing detail in dim astrophotographs, Photoshop only allows permanent adjustments to the histogram.

Histograms and Curves

All imaging processing programs have the ability to display a histogram of the image. This graph describes the distribution of brightness of the pixels in the image. The left-hand side is the black point, and the right edge is the white point. Any pixel at or near the black point will appear black, and any pixel at or near the white point will appear white. Your goal in processing is to create a beautiful image with a balanced tone and a wide-dynamic range, which usually has a bell-shaped histogram peaking in the left third of the scale. Ideally, there should be few purely black or purely white pixels, so the curve should be approach zero at both ends. Various programs scale the histogram differently, so become familiar with an ideal histogram for whatever program that you use (see Fig. 15.1).

The quickest way to create a balanced image is with digital development (DDP) in your astronomical software. DDP is designed to create an image similar to the response of film, by reducing the apparent differences in brightness between bright and dark areas on the image. DDP works its magic best with large nebulas or galaxies that have a broad brightness range, but poorly with open clusters that are basically bright stars and a dark background. Set the background level to prevent the black point from entering into the bell shape of the histogram curve, which could hide detail in the darker areas of the image. Similarly, set the mid-level to display detail in nebulosity and galactic disks without burning out the brighter areas of nebulosity or the cores of galaxies. View your histogram at maximum values for the screen stretch to see the full impact of these

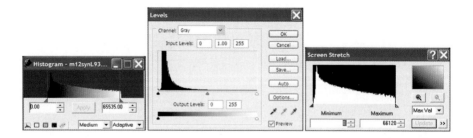

Fig. 15.1. Histogram displays in CCDSoft (*left*), Photoshop (*center*), and MaximDL (*right*) for a globular cluster after successful digital development in MaximDL. *Black point* is at *left* border of each graph, and *white point* is at the *right* edge of each graph.

The Order of Image Processing

effects. When in doubt, apply DDP gently to your image. You can finish adjusting brightness later with curves and levels.

Levels and Curves in Photoshop can yield finer control of contrast in your image than is possible using DDP. Many other programs provide similar capabilities by using a combination of adjustments of the white and black points of the histogram, followed by application of curves. The first step is the adjustment of the white point in the image. Bring the white point into the curve until it reaches the brightest pixels at the far right edge the histogram curve. This will define only the brightest pixels in the image as pure white. In some programs, such as Maxim DL, you will need to save this change as a histogram stretch. Then, in this revised histogram, move the black point until it reaches the left edge of the bell-shaped curve. Avoid forcing the black point into the upslope of the curve, which could obscure detail in the darker areas of the image. Third, apply a gentle curve to increase the brightness levels of the midrange of pixel values. Keep the right side of the curve straight to avoid excessive brightness in lighter areas of the image. Several applications of gentle curves provide better control of contrast than a single harsh curve. Because your curves change brightness levels, the peak of the bell-shaped histogram shifts away from the black point, so you will need to move the black point after the application of curves in order to keep the background sky dark.

Some imagers prefer to apply DDP or curves to individual channels before color combining, especially because DDP can reduce color saturation. However, if you apply histogram adjustments to each color channel separately, you run the risk of boosting different colors at different brightness levels, ruining color balance. If your RGB image lacks color contrast, you can restore color intensity in several ways, as discussed in the section "Color Enrichment."

Luminance Layering

The human eye sees details in the luminance (brightness) component of an image and not in the color component. Therefore, the color information of an image can be obtained with a lower resolution than the luminance data, without degrading the perceived quality of an image. Because color data can be acquired with lower resolution, the color exposures in an LRGB sequence can be binned 2×2, accumulating in 15 min the same signal-to-noise as a 1-h color exposure unbinned. A monochrome camera with a clear filter collects as much light in an hour as color filters collect in 4 h. For example, if you have 2 h for imaging, you can choose an LRGB technique devoting 60 min to luminance and 20 min each to red, green, and blue exposures. Compared with a conventional RGB sequence devoting 40 min each to red, green, and blue exposures, the 2-h LRGB method yields twice as much luminance signal (allowing sharper detail), and twice the color signal-to-noise, in the same time.

As a rough guideline, plan on spending half of your imaging time in luminance, and half in RGB. This ratio will change with very long imaging sequences, which benefit by devoting a higher percentage of imaging time to the luminance. For very dim objects, you can obtain your clear luminance exposures binned 2×2 and your RGB exposures binned 3×3.

For some objects, a filtered luminance can give better results. For example, a bright emission nebula like the Rosette will show richer contrast in the nebula if the luminance is acquired with a red filter. This can be approached in two ways. One choice is to acquire the red, green, and blue exposures unbinned, but with more exposure time given to red. For the standard 2-h exposure, you might devote 60 min to red, and 30 min each to green and blue. When processing, the red channel is used both for the luminance channel and for the red channel of the RGB data. A second choice is to obtain the red channel unbinned, and the green and blue channels binned 2×2. This would allow shorter green and blue exposures, so that more imaging time could be devoted to the red luminance. For a dim emission nebula like the California, a luminance using a hydrogen alpha (H-alpha) filter can yield more detail.

When you process your LRGB images, first perform image calibration. Then, modify the size of your RGB exposures to match the scale of your luminance exposures. Follow this by alignment of all images, choosing a good luminance image as the reference image. If your software does not allow you to align these many images, you can align your individual rescaled RGB exposures to the combined luminance exposure.

If your combined luminance channel has good signal to noise, perform deconvolution to enhance sharpness. Digital development (DDP) can be

The Order of Image Processing

performed on the deconvoluted luminance, but you should avoid any additional sharpening during digital development. If you prefer, you can perform histogram adjustments with levels and curves instead of DDP.

Combine your RGB channels using routine techniques (see the section "Color Combining"). Digital development can be performed on the combined RGB image, which will reduce color contrast, but helps to preserve color balance (see the section "Histograms and Curves"). Most astronomical software has provisions for LRGB combination, but you will get better results and control of your image by applying the luminance as a separate layer in Photoshop.

Open your luminance in Photoshop. If you have an old version of Photoshop that only works with 8-bit files, use a FITS plug-in (available free from http://fits.gsfc.nasa.gov/fits_viewer.html) to allow you to perform your final levels and curves adjustment in 16 bits. After you are satisfied with the contrast and brightness of your luminance image, convert the FITS file into an 8-bit TIFF file. If you have a more recent version of Photoshop, you can work with 16-bit TIFF images. To start, use your astronomical software to convert your luminance image from FITS to 16 bit TIFF. Open the luminance file in Photoshop, and adjust your levels and curves until you are satisfied with contrast and brightness.

Then, convert your RGB image into TIFF format and open it in Photoshop. Click on your luminance image, select all (Ctrl+A), copy (Ctrl+C), click back on your RGB image, and then paste (Ctrl+V) to create a new layer over your RGB image. Name your new layer "luminance." Confirm alignment by changing blend mode of the luminance layer to difference; stars should appear as targets. If there is any misalignment, either nudge the luminance layer to align, or go back to your astronomical software to align the RGB image to your luminance. After alignment is confirmed, you can convert your blending mode to luminance. Save your new file with LRGB in the title.

In some cases, the LRGB image will have strange colors. This is especially a problem with a filtered luminance, most commonly with an H-alpha luminance. Robert Gendler suggests that an H-alpha luminance can be applied several times, beginning with a low opacity and gradually increasing opacity levels, flattening the image each time to strengthen the RGB, to better match the luminance to the RGB. Another choice is to apply an H-alpha luminance with either a "screen" or "lighten" mode instead of "luminance." A third option is to blend the H-alpha luminance with the red channel, at a ratio of about 4:1 of H-alpha:Red, either in your astronomical software or in Photoshop. The resulting new HR blend can be used both for the red channel of the RGB image and for the luminance channel. The HR luminance is usually set to an opacity of 50% or less. This blended approach also better matches the small sizes of stars on H-alpha with the larger sizes on RGB images.

Color Enrichment

Luminance layering and DDP sometimes suppresses color contrast. This can be restored in three ways. The easiest is by first selecting the RGB layer, and then increasing saturation until color intensity is restored. Another approach, using the "match color" command in more recent versions of Photoshop, sometimes enhances color better while introducing less color noise than boosting saturation. A third method using the Lab color model boosts contrast between blue and yellow, which is nice for some galaxies. First, lightly boost color intensity in the RGB layer with either saturation or "match color," and adjust color balance as needed. Then, flatten the image. Convert image mode to Lab Color. In channels, select the "b" channel, which represents blue-yellow contrast, and then click on the small box to the left of Lab to show all channels. Click on image, adjustments, brightness/contrast, and move the contrast slider to the right until the desired effect is achieved, typically at "+20" to "+50." Then, convert the image mode back to RGB color.

The Order of Image Processing

Image Sharpening and Blurring

Bright areas of an image have enough signal to benefit from sharpening. Dark areas of an image have weak signal, giving rise to a grainy appearance that could benefit from blurring. Most astronomical image programs have the filters to both sharpen images and blur images. The disadvantage of these routines is that they usually apply to the entire image. Photoshop has the ability to select a region of the image to apply the filter selectively.

Sharpening is usually reserved for the luminance channel of an image. If you are working with an RGB image, you can "fade" the filter to the luminance by clicking on edit, fade, luminance, 100% (or less) to limit the filter to the luminance portion of an RGB image. This avoids sharpening color, which can create a grainy color texture. If your image has a luminance with a low opacity under 50%, you can first flatten the image before sharpening, and then fade the filter to luminance.

Use the lasso tool to select the brighter areas of the image that you wish to sharpen. Then, use the magic wand tool to deselect (Alt click) the bright stars in your selection. Finally, shrink your selection by 5 pixels (select, modify, contract) and feather your border by 5 pixels (select, feather). Your selection is now ready to apply sharpening.

The easiest sharpening tool is "unsharp masking." Typically, choose an amount between 50 and 100%, a radius of 3–5 pixels, and a threshold of 0 if you have a good image, but a higher threshold if your image is noisy. A more precise sharpening tool is "smart sharpen," that is only available on more recent versions of Photoshop. As with unsharp masking, choose an amount of 50–100% and a radius of about 5 pixels. If your sharpening effect is too harsh, fade the application of sharpening to less than 100%. If your sharpening was on an RGB image, fade the blend mode to luminance.

Some astronomers prefer to use a high-pass filter for sharpening. In Photoshop, this must be applied to a duplicate of the layer that you wish to sharpen. Select the region that you wish to sharpen on the duplicate layer. Apply the high-pass filter to this duplicate layer (filter > other > high pass) with a radius of about 5, or the same as amount of pixels that you used for contract and feather in your selection. Then, change the blend mode for this layer to overlay. If the sharpening effect seems too harsh, you can reduce the opacity of this layer to less than 100%.

Blurring seeks to reduce image noise at the edges of the target and create a smooth background sky. Because noise has both luminance and color components, the image should be flattened before applying noise reduction.

In Photoshop, begin by selecting the background areas that you wish to smooth. This can be accomplished either with the magic wand tool or with the color range tool selecting shadows. If you have a recent version of Photoshop, the "reduce noise" filter works very well. Begin with the default values but adjust the various options depending on the severity of your noise and the amount of color noise. If you have an older version of Photoshop, you can use the Gaussian blur filter, but this can erase smaller stars. To reduce the effect on small stars, contract your background selection by 3 pixels and feather by 2 pixels, then apply a Gaussian blur of not more than 2 pixels.

Dealing out Gradients

Before attacking gradients, perform any final histogram and color adjustments until you are satisfied with the image contrast and brightness in your image. Many options are available to dispense with gradients. You may want to first try gradient reduction tools present in many astronomical imaging programs.

Within Photoshop, you can use the image itself to create a model of the sky gradient if the celestial target is either an open cluster or a well-defined nebula or galaxy that occupies less then a quarter of the image. Flatten the image (if not already flattened) and then duplicate the background layer. Choose the copy, and select color range – shadows. Select inverse, and then cut the selection. Apply a large Gaussian blur of about 50 or more pixels. Change the blend mode to difference, and reduce the opacity to about 80%. Your sky gradient should be almost invisible. This technique fails with larger faint nebula or galaxies because the faint detail can be erased and gradients superimposed directly on the target are not eliminated.

Other methods in Photoshop work for larger objects, using the gradient tool. This technique begins with a new blank layer that is "filled" with a gradient from sampled "foreground" in the brightest corner of your image to sampled "background" in the opposite corner. This gradient is then subtracted by changing the blending mode to difference at less than 100% opacity.

Another way to reduce gradients is with GradientXTerminator, which is a plug-in for Photoshop. This program creates a higher order correction to the background than can be achieved by simpler Photoshop routines and only takes a few minutes to perform. A 30-day trial of the software is available prior to purchase. The program includes a straightforward online tutorial.

Final Cleanup

Final processing steps include removal of any remaining hot and cold pixels, correction of excessive halos around bloated stars, and cropping the images. Zoom into your image at 200% or 300% to identify bad pixels. Many of these can be erased quickly using the spot healing brush tool in Photoshop. If you are using another software program or an earlier version of Photoshop, try the clone stamp tool to sample an adjacent area of normal signal and then paste over the artifact. Adjust the tool to the size of your artifacts, and keep hardness low to create a smooth transition. For repair of residual blooming spikes, try the clone stamp with an opacity of under 100% to gradually conceal the spike.

Bloated stars and halos can be corrected in many ways. If you have just a few bloated stars, begin with the lasso tool to select a generous area around the worst offending star and halo. Select additional stars by holding the shift key while employing the lasso tool. Feather your selection by a few pixels. Then use the minimum filter (filter, other, minimum) at a setting of 1 or 2, which will shrink both the halo and star. If the effect appears too harsh, you can fade the effect (edit, fade minimum filter) to less than 100%. If you have many halos to correct, you can select most of the halos at once using the color range command (select, color range) with the eyedropper tool. Expand your selection by a few pixels, and then feather your selection. Apply the minimum filter at a setting of 1, and then fade the filter if necessary. If the halos have an odd green tone, you can reduce their saturation (image, adjustments, and hue/saturation).

Image alignment leaves dark borders around many images, which creates a noisy edge to your final image. Initial cropping should eliminate this border. You may perform additional cropping to create a more balanced image. Avoid cropping to just place your target in the center of the image. Many images will appear more dynamic with the main target off center, especially if the target is balanced by some bright stars, nebulosity, or background galaxies in the other side of the image.

About the Author

Ruben Kier grew up under dark suburban skies in North Carolina where almost every kid in the neighborhood had a small telescope. He began his formal education in astronomy at Cornell University in 1974, when Carl Sagan was their faculty. He continued studying the sciences, graduating Harvard University with highest honors in Biochemistry. After medical training at Duke, he joined the Yale faculty, and went on to publish over 50 scientific articles and chapters in academic radiology, becoming associate editor of the journal *Radiology*.

His interest in astronomy was rekindled with his daughters, who shared his excitement during the visits of comets Hyakutake and Hale-Bopp. In 1998, daughter Melanie won first place in the children's division of Astronomy Magazine's 25th Anniversary Photo competition, and the next year daughter Shelley had a letter published in *Astronomy* magazine. Since then, Dr. Kier has had several articles, letters, and numerous photos published in *Astronomy* and *Sky & Telescope* magazines. His goal has been to show how amateur astronomers, using moderately priced equipment, can obtain excellent images rivaling those obtained with professional telescopes.

About the Author

The author and his Meade 12-in. telescope at the Hidden Lake Observatory.

Index

A
Abell 21, 60
Aligning, 342
Andromeda Galaxy, 262
Angel Nebula, 42
Antiblooming, 332
Aperture mask, 329
Arp Peculiar Galaxy
    A214, 105
    A316, 90
    A317, 99, 102
    A319, 243
Astro-shed, 318
Atmospheric extinction, 322
Atmospheric refraction, 322
Atmospheric seeing, 324
Autoguiding, 313, 330

B
Bias frame, 334
Bloated stars, 354
Blue Snowball Nebula, 258
Bubble Nebula, 255

C
Caldwell
    C4, 223
    C5, 4
    C6, 172
    C7, 71
    C9, 252
    C11, 255
    C12, 207
    C13, 274
    C14, 290
    C19, 234
    C20, 217
    C22, 258
    C23, 292
    C26, 113
    C27, 204
    C30, 246
    C31, 16
    C32, 131
    C33, 209, 211
    C34, 209
    C38, 125
    C39, 63
    C49, 51
    C50, 51
    C56, 265
    C57, 198
    C63, 240
    C65, 268
Calibration, 340
California Nebula, 10

Index

Cat's Eye Nebula, 172
Cave Nebula, 252
Cepheus Flare, 237
Christmas Tree Cluster, 54
Cigar Galaxy, 82
Clownface Nebula, 63
Cocoon Nebula, 234
Color combine, 345
Composition, 327
Cone Nebula, 54
Crab Nebula, 25
Crescent Nebula, 204
Curves and levels, 347

D
Dark frame, 334
Deblooming, 341
Deconvolution, 344
Deer Lick Galaxy Group, 246
Digital development, 346
Double Cluster, 290
Drift alignment, 321
Dumbbell Nebula, 201

E
Eagle Nebula, 181
Elephant's Trunk, 229
Emission nebula, xx
Eskimo Nebula, 63
Exposing, 332

F
Fetus Nebula, 220
Fireworks Galaxy, 207
FITS plug-in, 349
Flame Nebula, 31
Flaming Star Nebula, 16, 18
Flat field, 336
Flying Horse Nebula, 249
Focusing, 328
Framing, 322, 326

G
Galaxy, xxi
Globular cluster, xx, 167
Gradients, 353
Great Hercules Cluster, 164

H
Hamburger Galaxy, 102
Heart Nebula, 298, 299
Helix Nebula, 240
Hickson 44 Galaxy Group, 90

Hidden Treasure
  T3, 271
  T8, 287
  T32, 28
  T34, 31
  T38, 54
  T51, 77
  T52, 87
  T58, 102
  T67, 131
  T69, 134
  T103, 220
  T105, 229
  T106, 249
Histogram, 346
Hockey Stick Galaxy, 131
Horsehead Nebula, 31

I
IC 342, 4
IC 405, 16
IC 410, 16, 19
IC 434, 31
IC 443, 48
IC 1396, 229
IC 1805, 298
IC 1848, 298
IC 2118, 13
IC 5070, 214
IC 5146, 234
Iris Nebula, 223

J
Jellyfish Nebula, 48

L
Lagoon Nebula, 175, 178
Leo Trio, 99
Little Dumbbell Nebula, 284
Little Pinwheel Galaxy, 87
Luminance layering, 348

M
Median combine, 343
Medusa Nebula, 60
Meridian, 322
Messier
  M1, 25
  M2, 226
  M3, 150
  M5, 159
  M8, 175, 178
  M11, 190